$24 95

RELICS OF EDEN

Daniel J. Fairbanks

RELICS OF EDEN

THE POWERFUL EVIDENCE OF EVOLUTION IN HUMAN DNA

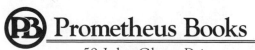

Prometheus Books

59 John Glenn Drive
Amherst, New York 14228–2119

Published 2007 by Prometheus Books

Inquiries should be addressed to
Prometheus Books
59 John Glenn Drive
Amherst, New York 14228–2119
VOICE: 716–691–0133, ext. 210
FAX: 716–691–0137
WWW.PROMETHEUSBOOKS.COM

11 10 09 08 07   5 4 3 2 1

Library of Congress Cataloging-in-Publication Data

Fairbanks, Daniel J.
 Relics of Eden : the powerful evidence of evolution in human DNA /
Daniel J. Fairbanks.
      p. cm.
 Includes bibliographical references and index.
 ISBN 978–1–59102–564–1
 1. Molecular evolution. 2. Evolutionary genetics. I. Title.
[DNLM 1. Evolution, Molecular. 2. Genome, Human. 3. Genomics—history.
4. Human Genome Project. QU 475 F164r 2007]

QH390.F35 2007
572.8'38—dc22

2007027126

Printed in the United States of America on acid-free paper

CONTENTS

6 CONTENTS

PREFACE

In recent years, advocates of the creationism and intelligent design movements have successfully promoted history's most sophisticated and generously funded attack on science, claiming that evolution, human evolution in particular, is "a theory in crisis." Ironically, during this same period, the Human Genome Project revealed the most powerful evidence of human evolution ever discovered, and other genome projects, especially the chimpanzee and rhesus macaque projects, have substantially augmented that evidence.

With the ongoing controversy over intelligent design, people often ask me to recommend a book on the molecular evidence of human evolution. Unfortunately, most popular human evolution books either fail to include DNA evidence, or, if they do, they cover only a few highlights. Instead, they tend to focus on archaeological, geological, anatomical, physiological, and theoretical evidence with little or no discussion of the literally millions of molecular fossils in DNA. These requests, and my dismay at repeated claims of meager and flawed evidence supporting human evolution, led me to draft the book you are now reading.

This book comes at a time when the clash between those who perceive evolution as a threat to religious faith and those who view religious fervor as a threat to science is as intense as it ever has been. The first eight chapters of this book, and the three appendices, focus on pure and solid science. Toward the end, however, the final two chapters address this clash and argue for the middle ground—that science and religion are complementary ways of seeking answers and they need not be at odds.

Most of you reading this have varied backgrounds and interests in science, especially in molecular biology. To satisfy these diverse needs, I have chosen to keep the main text of the book as simple and as concise as possible, while still telling the remarkable story the evidence has to offer. For those of you who want additional information, I have written two appendices that explore some of the main topics in the book in more detail and a third appendix that tells the story of how some of history's best scientific minds laid the foundation for the discoveries presented in this book.

You will shortly read about evidence discovered by hundreds of scientists. I wish first to acknowledge them and their work, and I hope this book does justice to their extraordinary accomplishments. I also wish to acknowledge those who taught me science, the first being my grandfather, Avard T. Fairbanks, who as a professor, human anatomist, and prominent sculptor opened my eyes to the magnificence of both science and art. I also am especially grateful to professors Clayton White and Duane Jeffery, who, when I was an undergraduate student, helped me explore the wonders of evolution. I recorded the very day, January 30, 1979, when during a discussion on evolution in Dr. White's class I discovered a newfound way of thinking as a scientist. Shortly thereafter I met Dr. Jeffery, who personally helped me work through the change in thought required by that discovery. Now, twenty-six years later, his review of the manuscript for this book has been invaluable.

I owe special thanks to Cecie Starr, who introduced me to book authorship. I greatly appreciate her extraordinary encouragement and assistance over the years and during the development of this book. The teachings of many other professors and colleagues have inspired my passion for science and are reflected in the pages that follow. To them I likewise am grateful. I also thank the thousands of students who enliven the classes I teach. There is no better profession than being a university professor, and the students make it all worthwhile.

Throughout the publication process, the editors and staff of Prometheus Books have exemplified the utmost courtesy and professionalism. I applaud them for their superb work.

It should go without saying that the opinions in this book are entirely my own, and in expressing them by no means do I intend to represent the views of my employer, my colleagues, or the publisher. I have scrupulously tried to ensure that there are no errors but have learned from past publishing experience that in spite of my best efforts at eradicating errors, a few inevitably persist. I apologize in advance for any errors and take sole responsibility for them.

Finally, to my wife, Dr. Donna Fairbanks, and my children, Jonathan, Aaron, and Michael, I am indebted to you for your patience and support.

Introduction

FOSSILS IN OUR DNA

A wise professor of Russian literature once told me that every university student should take a course in geology. With my love of science, I needed little convincing. It was the spring of 1982, and in a university with over twenty-five thousand students, only three of us enrolled in the course. (To this day I'm still baffled by the lack of interest most students show toward science.) We were the luckiest three students on campus that term. With such a small class, the professor canceled the classroom and we spent most of our time traveling through the mountains studying geological formations. I was astonished at how often we found fossils. For the last week of class, we camped in Capitol Reef National Park, one of the most geologically rich places on Earth. Our professor hadn't told us how he would determine our grades, and we never had any tests, just constant verbal quizzing from him. As we were leaving the park, he pulled the car to the side of the road and told us to climb a nearby hill. "Anyone who finds a fossil gets an A in the class," he said with a smirk on his face. We dutifully scampered up the hill to find ourselves on a plateau covered with thousands of fossilized oyster shells. We all got our As.

That summer I saw abundant evidence that the earth was very old and that it contains countless remnants of long-extinct organisms. I saw how scientists carefully studied the multitude of fossils they found to reconstruct as much as possible the history of life on Earth. The fossils we saw that summer in very ancient rocks were obviously different from any modern organisms; those from more recent rocks looked much more like the animals and plants we see today. The progression of life over time from the strange to the familiar was readily apparent.

I then went to graduate school to study genetics and DNA, where I encountered fossils of a different kind: relics buried in the DNA of all organisms, including ours. Those relics tell a fascinating story. To understand how the story goes, let's start with a simple analogy. As often happens in our electronic age, I received not long ago an amusing e-mail message from a friend. This particular message had a list of predicted headlines from the year 2029; "Baby Conceived Naturally, Scientists Stumped," read one of them. My friend received it from a friend, who received it from a friend, and so on. Before e-mail, people passed on messages like these as photocopied pages. So let's think back to those days and imagine someone photocopying a page with an amusing message and passing it on to several friends, who photocopy their copies and pass them on to their friends who do the same, and so on. Each time someone photocopies the message, specks of dust on the photocopier glass end up leaving marks on the new copy. These marks are preserved on subsequent copies every time that particular page, or a copy of that page, is photocopied. New marks are added with each round of photocopying so that these extraneous marks end up accumulating until the copies are cluttered with them. Each mark is a relic added at some point in the branching chain of photocopying.

Now, imagine a good detective who gathers the thousands of copies of this message passed along through photocopying from friend to friend. This detective could reconstruct the branching chain of photocopying by comparing all of the pages and determining how many of the extraneous marks are the same and how many are different. Marks present on a large proportion of the pages must have occurred early in the chain; those on just a few happened later. Furthermore, this detective can hierarchically group the pages to reconstruct the branching chain purely by comparing similar and different marks on the pages, without any information about who had the pages or who received a copy from whom. These marks are relics of photocopying that tell a story about how the message on the page was passed on, even though they have nothing to do with the message itself.

Organisms replicate their DNA and pass it on to their offspring in a manner somewhat like this simple photocopying analogy, although the process of inheritance is much more complex. At times, useless bits of DNA creep into the mass of useful information and end up being passed on from one generation to the next. These relics of DNA accumulate generation after generation until there are literally millions of them. Although most of them have no relationship to the message in DNA, cells do not routinely purge them; instead they copy them, over and over, as the cells divide to form more cells. These relics, in a sense, are fossils buried in the DNA of every organism on Earth, waiting for us to uncover them. Like our hypothetical detective, scientists who examine them can decipher the stories they tell about genetic histories.

By the time I received my PhD in 1988, studies of DNA had produced powerful evidence of evolution, as powerful as the overwhelming evidence from fossils in rocks. In the ensuing years, the evidence from DNA grew at a rate so astounding that even the most optimistic scientists were stunned. The most massive evidence came from genome projects, especially the Human Genome Project, which deposited enormous amounts of DNA sequence information in computer databases. The Internet made those sequences easily and freely available to anyone in the world with an Internet connection. The opportunity to search DNA for the story of life was never greater. What once was a trickling stream of DNA-based studies on human evolution became a torrent. One study after another showed that our DNA is littered with literally millions of relics of our evolutionary ancestry.

In 1998, the National Academy of Sciences, the most prestigious and highly respected group of scientists in the United States, issued a report with this concise and exceptionally well-worded statement: "It is no longer possible to sustain scientifically the view that living things did not evolve from earlier forms or that the human species was not produced by the same evolutionary mechanisms that apply to the rest of the living world."[1] Few biologists question the reality of evolution. The same cannot be said of the general public. Sadly, few people understand modern evolutionary theory and the scientific evidence on which it is founded. Moreover, few have studied the social culture that has fueled a clash between religious beliefs and evolutionary science for more than one hundred and fifty years. For some, scientific ignorance is bliss. But in today's world, where science plays such a vital role in society, a better understanding of evolution is needed.

The intent of this book is to present just a fraction, but a very compelling fraction, of the DNA-based evidence of evolution. I have chosen to focus on human evolution because some people are willing to accept the idea that other species have evolved but draw the line with humans, usually for religious reasons. Yet thanks to the Human Genome Project, we now have more evidence of evolution for humans than for any other species. Lest anyone think that my purpose in writing this book is to criticize religion, be assured it is not. Like many of my scientific colleagues, I hold deep religious convictions. However, along with them, I strongly believe that attempts to discredit the powerful evidence of evolution actually *harm* faith rather than promote it.

We'll return to the debate over evolution and religion at the end of this book. For now, we have a fascinating journey ahead of us as we explore some of the many relics of evolution found in the DNA of every person on the planet.

NOTE

1. National Academy of Sciences, *Teaching about Evolution and the Nature of Science* (Washington, DC: National Academy Press, 1998), p. 16.

Chapter 1

FUSION

Images of evolution abound, often depicting humans with apes and monkeys. The idea that we are related to other primates gripped the public almost from the day Darwin first published the *Origin of Species*. On June 30, 1860, just a few months after the first edition of *Origin of Species* was released, Samuel Wilberforce, the bishop of Oxford, and Thomas Henry Huxley, a staunch supporter of Darwin, engaged in a brisk exchange during a meeting at Oxford University. Years later, Isabella Sidgwick famously recounted the incident with more than a bit of embellishment:

> The Bishop rose, and in a light scoffing tone, florid and fluent, he assured us there was nothing in the idea of evolution; rock-pigeons were what rock-pigeons had always been. Then, turning to his antagonist with a smiling insolence, he begged to know, was it through his grandfather or his grandmother that he claimed his descent from a monkey? On this Mr. Huxley slowly and deliberately arose. A slight tall figure stern and pale, very quiet and very grave, he stood before us, and spoke those tremendous words—words which no one seems

sure of now, nor I think, could remember just after they were spoken, for their meaning took away our breath, though it left us in no doubt as to what it was. He was not ashamed to have a monkey for his ancestor; but he would be ashamed to be connected with a man who used great gifts to obscure the truth. No one doubted his meaning and the effect was tremendous. One lady fainted and had to be carried out: I, for one, jumped out of my seat.[1]

Huxley's own recollection, although less spirited than Ms. Sidgwick's, captured the excitement of the moment:

If then, said I, the question is put to me would I rather have a miserable ape for a grandfather or a man highly endowed by nature and possessed of great means of influence and yet who employs those faculties and that influence for the mere purpose of introducing ridicule into a grave scientific discussion —I unhesitatingly affirm my preference for the ape. Whereupon there was inextinguishable laughter among the people —and they listened to the rest of my argument with the greatest attention.[2]

Almost a century and a half later, emotions still run high whether one accepts or scorns the idea of an evolutionary relationship between humans and apes. Setting emotions aside, what does science tell us about the relationship? Specifically, does our DNA hold solid evidence that humans and great apes share common ancestry? This chapter highlights some of the clearest and most powerful evidence, all within a single human chromosome.

HUMAN AND GREAT-APE CHROMOSOMES
ARE STRIKINGLY SIMILAR.

The DNA of most organisms, including humans, is contained within structures called *chromosomes*. Each chromosome is a coiled-up molecule of DNA with proteins that stabilize it. When a cell is about to divide in two, the chromosomes condense into compact structures that are easy to observe with the help of a microscope. Each compact chromosome has a constricted region called a *centromere*, with very specific DNA sequences in it. The centromere serves a critical role in helping direct the chromosome to its proper position when cells divide. The ends of each chromosome are called *telomeres* and they, too, contain highly specific DNA sequences.

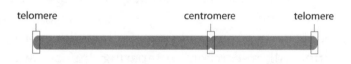

Each time a chromosome replicates, a bit of telomere DNA erodes away, but a protein called telomerase restores the eroded ends to reconstitute the telomeres. Thus, telomeres function as buffer zones to protect the important DNA within the chromosome from erosion. If not for telomeres and telomerase, our chromosomes would progressively erode inward from the ends until they could no longer function. As we'll see momentarily, these structural features—centromeres and telomeres—offer a compelling clue about our evolutionary history.

If we compare human chromosomes to those of the great apes (chimpanzee, gorilla, and orangutan), one glaring difference stands out: the human genome has one fewer chromosome than the genomes of the great apes. Chimpanzee, gorilla, and orangutan genomes each have twenty-four chromosomes, whereas the human genome has twenty-three. Some opponents of evolution pounced on this difference, claiming that it disproves common ancestry between humans and apes. At one time the difference seemed to justify the traditional classification of great apes into a single family called the Pongidae and humans as the sole surviving species of a different family called the Hominidae. However, DNA analysis eventually shattered that classification (as we'll see later).

In 1982, Jorge Yunis and Om Prakash published what is now viewed as a landmark article in the journal *Science*.[3] Their work confirmed the findings of several earlier studies in exquisite detail: chromosomes from humans, chimpanzees, gorillas, and orangutans are highly similar and can be aligned with one another. More recently, the human and chimpanzee genome projects confirmed and greatly amplified this comparison, showing that not just the chromosomes but the DNA within the chromosomes of humans and chimpanzees is strikingly similar in both sequence and organization, usually with at least 98 percent identity. All researchers comparing human and chimpanzee chromosomes, and the DNA in them, arrived at the same overall conclusion: every human chromosome, except one, has a matching chimpanzee chromosome. The exception is human chromosome 2, which matches *two* different chromosomes in chimpanzees as well as in the other great apes.

HOW DID HUMAN CHROMOSOME 2 ORIGINATE?

The two chimpanzee chromosomes that match human chromosome 2 are called 2A and 2B, and their structures and DNA sequences align them along their full lengths with human chromosome 2.

What could explain this curious alignment of two chimpanzee chromosomes with one human chromosome? There are three mutually exclusive possibilities.

First, if humans and chimpanzees share a common ancestor, human chromosome 2 might have formed by the fusion of two chromosomes after the lineage leading to modern humans split from the one leading to modern chimpanzees.

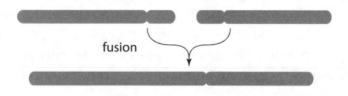

Second, chimpanzee chromosomes 2A and 2B might have resulted from fission (breakage) of an ancestral chromosome into two after the lineages leading to humans and chimpanzees split.

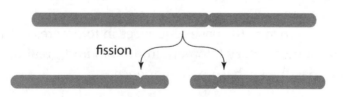

Third, the human and chimpanzee chromosomes may have originated independently with no evolutionary relationship to each other. In this case, the similarity is not a product of common ancestry, merely an illusion of it.

Nine years after Yunis and Prakash published their article,

DNA analysis resolved the dilemma, unambiguously supporting one of these three possibilities. To understand how it did, we need a little background information about DNA. DNA molecules contain strands of bases, molecular units that are repeated one after another. There are four different bases in DNA, each designated by the letter representing its chemical name: T, C, A, and G. For example, we can represent a single strand of DNA as a series of letters corresponding to the following base sequence:

ATGGTGCACCTGACTCCTGAGGAGAA

Most DNA molecules contain two strands, which are paired with each other. Wherever there is a T in one strand, an A is paired with it in the other. Wherever there is a C in one strand, a G is paired with it in the other. For example, part of the base sequences in two paired strands of DNA can be written as

ATGGTGCACCTGACTCCTGAGGAGAA
TACCACGTGGACTGAGGACTCCTCTT

Every T in one strand is paired with A in the other, and the same is true for G and C.

Now let's turn to the base sequences in telomeres at the ends of chromosomes. Every telomere in human and great-ape chromosomes has the six base-pair sequence

TTAGGG
AATCCC

repeated over and over about fifty to one hundred times in tandem:

```
...TTAGGG|TTAGGG|TTAGGG|TTAGGG|TTAGGG|TTAGGG|TTAGGG|...
...AATCCC|AATCCC|AATCCC|AATCCC|AATCCC|AATCCC|AATCCC|...
```

In 1991, scientists at Yale University sequenced the DNA from the site in the middle of human chromosome 2 that matches telomeres at the ends of chimpanzee chromosomes 2A and 2B.[4]

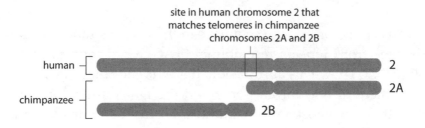

Their findings clearly reveal which of the three explanations is best. In this region is a DNA sequence with 158 copies of the tandem repeat found in telomeres. At some point in human ancestry, the telomere of one chromosome fused head-to-head with the telomere of a different chromosome. The exact fusion site is preserved in the DNA, and figure 1.1 (p. 25) shows all 158 copies of the sequence with the fusion site labeled. The massive sequence in figure 1.1 is a bit daunting, so let's focus just on the fusion site.

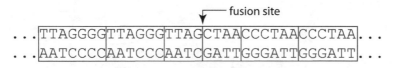

To understand how this sequence reveals the exact fusion site, we need to recognize that the two strands of a DNA molecule do not have a "right side up." The repeat found in telomeres is

```
                    TTAGGG
                    AATCCC
```

However, if we rotate the above sequence 180 degrees, it reads

```
                    CCCTAA
                    GGGATT
```

At the fusion site the sequence in the upper strand abruptly changes from repeats resembling TTAGGG to repeats resembling CCCTAA. This abrupt switch is evidence that the DNA in a telomere of one chromosome and the DNA in a telomere of the other chromosome broke and then the two chromosomes fused at the broken ends. Based on evidence of this past event and on what we know about how chromosomes break and fuse in modern times, here's how the fusion process probably took place:

First, the two separate chromosomes had identical telomere sequences with multiple tandem repeats of TTAGGG.

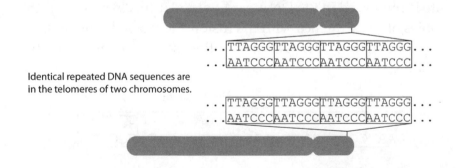

Identical repeated DNA sequences are in the telomeres of two chromosomes.

Both chromosomes broke within the telomeres, leaving four base pairs of what was a six base-pair repeat at their broken ends.

```
TTATCCCAAAGCAAGGCGAGGGGCTGCATTGCAGGG TGAGGG TGAGGG TGAGGG TGAGGG
AATAGGGTTTCGTTCCGCTCCCCGACGTAACGTCCC ACACCC ACTCCC ACTCCC ACTCCC
```

```
TTAGGG TTTGGG TTGGGG TTGGGG TTGGGG TTGGGG TAGGG TTGGGG TTTGGG TTGGGG
AATCCC AAACCC AACCCC AACCCC AACCCC AACCCC ATCCC AACCCC AAACCC AACCCC
```

```
TTAGGG TTAGGGG TAGGGG TAGGG TCAGGG TCAGGG TCAGGG TTAGGG TTTTAGGG TTAGGG
AATCCC AATCCCC ATCCCC ATCCC AGTCCC AGTCCC AGTCCC AATCCC AAAATCCC AATCCC
```

```
TTAGGG TTAAGG TTTGGG TTGGGG TTGGGG TTGGGG TTAGGGG TTAGGGG TTAGGGG
AATCCC AATTCC AAACCC AACCCC AACCCC AACCCC AATCCCC AATCCCC AATCCCC
```

```
TTAGGG TTGGGG TTGGGGG TTGGGG TTGGGG TTAGGG TAGGGG TAGGGG TAGGG
AATCCC AACCCC AACCCCC AACCCC AACCCC AATCCC CATCCC CATCCC CATCCC
```

```
TTAGGG TTAGGG TTAGGG TAAGGG TTAAGGG TTGGGG TTGGGG TTGGGG TTAGGG
AATCCC AATCCC AATCCC ATTCCC AATTCCC AACCCC AACCCC AACCCC AATCCC
```

```
                  ┌── fusion site
TTAGGGG TTAGGGG TTAG CTAA CCCTAA CCCTAA CCCCTAA CCCCTAA CCCCAA CCCAAA CCCCAA
AATCCCC AATCCC AATC GATT GGGATT GGGATT GGGGATT GGGGATT GGGGTT GGGTTT GGGGTT
```

```
CCCCAA CCCCAA CCCTA CCCCTA CCCCTAA CCCCAA CCCTTAA CCCTTAA CCCTTAA CCCCTTA
GGGGTT GGGGTT GGGAT GGGGAT GGGGATT GGGGTT GGGAATT GGGAATT GGGAATT GGGAAT
```

```
CCCTAA CCCTAA CCCAAA CCCTAA CCCTAA CCCTA CCCTAA CCCAA CCCTAA CCCTAA CCCTA
GGGATT GGGATT GGGTTT GGGATT GGGATT GGGAT GGGATT GGGTT GGGATT GGGATT GGGAT
```

```
CCCTAA GCCTAAAA CCCTAAAA CCGTGA CCCTGA CCTTGA CCCTGA CCCTTAA CCCTTAA
GGGATT CGGATTTT GGGATTTT GGCACT GGGACT GGAACT GGGACT GGGAATT GGGAATT
```

```
CCCTTAA CCCTAA CCCTAA CCATAA CCCTAAA CCCTAA CCCTAAA CCCTAA CCCTA CCCTAA
GGGAATT GGGATT GGGATT GGTATT GGGATTT GGGATT GGGATTT GGGATT GGGAT GGGATT
```

```
CCCCAA CCCCTAA CCCTAA CCCCTATA CCCTAA CCCTAA CCCTA CCCCTA CCCCTAA
GGGGTT GGGGATT GGGATT GGGGATAT GGGATT GGGATT GGGAT GGGGAT GGGGATT
```

```
CCCCAA CCCCAGC CCCCAA CCCCAA CCCTTA CCCTAA CCCTA CCTAA CCCTTAA CCCTAA
GGGGTT GGGGTCG GGGGTT GGGGTT GGGAAT GGGATT GGGAT GGATT GGGAATT GGGATT
```

```
CCCCTAA CCCTAA CCCCTAA CCCTA CCCCAA CCCCAAA CCCAA CCCTAA CCCAA CCCTAA
GGGGATT GGGATT GGGGATT GGGAT GGGGTT GGGGTTT GGGTT GGGATT GGGTT GGGATT
```

```
CCCAA CCCTAA CCCCTA CCCTAA CCCCTAA CCCTAA CCCCTA CCCTAA CCCCTAA CCCTAA
GGGTT GGGATT GGGGAT GGGATT GGGGATT GGGATT GGGGAT GGGAT GGGGATT GGGATT
```

```
CCCCTA CCCTAA CCCCTAA CCCTAG CCCTAG CCCTAA CCCTAA CCCTCA CCCTAA CCCTCA
GGGGAT GGGATT GGGGATT GGGATC GGGATC GGGATT GGGATT GGGAGT GGGATT GGGAGT
```

```
CCCTAA CCCTCA CCCTCA CCCTCA CCCTCA CCCTAA CCCAA CGTCTGTGCTGAGAAGAAT
GGGATT GGGAGT GGGAGT GGGAGT GGGAGT GGGATT GGGTT GCAGACACGACTCTTCTTA
```

Figure 1.1. The DNA sequence from the site where two chromosomes fused to form human chromosome 2. The arrow indicates the exact fusion site. Each of the 158 repeated telomere sequences is boxed.

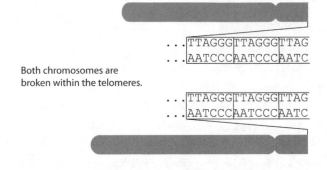

Both chromosomes are
broken within the telomeres.

The broken ends were ready to fuse with each other, which we can visualize if we rotate one of the chromosomes 180 degrees relative to the other, thereby inverting the DNA sequence.

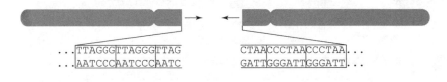

Once the two chromosomes have fused, the DNA sequence matches the fusion site in human chromosome 2.

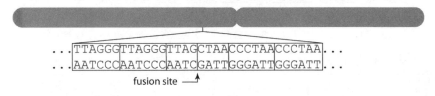

fusion site ⟶

Chromosome fusion is a very rare event. In fact, one of the functions of intact telomeres is to protect the ends of chromosomes from fusing with each other. However, when two chromosomes break, the protection against fusion is lost and they are free to fuse at the breakage points. There are several documented examples of recent chromosome fusions in humans, one of which is associated with a rare type of Down syndrome.

The proportion of people who have recent chromosome fusions is extremely small, but it provides clear evidence that new fusions do occur.

Let's now return to the repeated sequence surrounding the fusion site in figure 1.1. Of the 158 repeats, 44 are perfect copies of TTAGGG or CCCTAA. In most cases, the remaining repeats differ from the standard sequence by no more than one or two base pairs.

This is precisely what we expect if the fusion happened long ago in the remote ancestry of humans. After the fusion event, the repeats no longer functioned as telomeres, so mutations (changes in the DNA sequence) in them had no harmful or beneficial effect. The ancient telomere at the fusion site is now a nonfunctioning relic of evolution embedded in the middle of the chromosome. The more generations humans are separated from the fusion event, the more mutations we expect to accumulate in the sequence. Because the majority of the repeated segments have mutations in them, the chromosomes must have fused a long time ago, probably tens of thousands of generations deep into our ancestry. Thus, the evidence clearly eliminates chromosome fission and independent origins as reasonable alternatives to fusion.

THERE'S STILL MORE EVIDENCE OF FUSION.

These leftover telomere sequences in the middle of chromosome 2 are not the only evidence that this chromosome arose from a fusion. The centromere in human chromosome 2 aligns with the centromere in chimpanzee chromosome 2A, but the centromere in chimpanzee chromosome 2B aligns with a site in human chromosome 2 where there is no centromere.

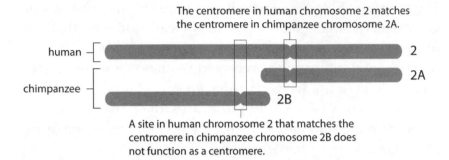

The centromere in human chromosome 2 matches the centromere in chimpanzee chromosome 2A.

human — 2
chimpanzee — 2A
2B

A site in human chromosome 2 that matches the centromere in chimpanzee chromosome 2B does not function as a centromere.

Fusion at the telomeres should have left *two* centromeres in the ancient fused chromosome, but there is only one now. Is there any evidence that a centromere was once present at this second site?

In every human and great-ape chromosome, the centromere contains a very specific DNA sequence that is repeated over and over, a 171 base-pair sequence called the alphoid sequence. Two groups of scientists, one in Italy and the other in the United States, searched for alphoid sequences in human chromosomes and found them at every centromere, as expected. They also found alphoid sequences at the site in human chromosome 2 where the remnants of this second centromere should be. These remnants are evidence of a now-defunct centromere.[5]

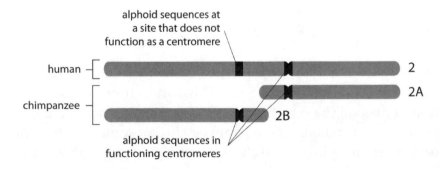

alphoid sequences at a site that does not function as a centromere

human — 2
chimpanzee — 2A
2B

alphoid sequences in functioning centromeres

All copies of the alphoid sequence in this nonfunctioning remnant are mutated when compared to those in functioning centromeres, evidence that this second centromere was disabled long ago in our ancestry, probably at around the time when the two chromosomes fused.[6]

The evidence that human chromosome 2 arose from a fusion, and that it closely matches two chimpanzee chromosomes (as well as two chromosomes in other great apes), is solid and unmistakable. The only reasonable explanation of this evidence is a chromosome fusion that happened after the lineage leading to humans diverged from the lineages leading to the great apes. This solid evidence of our common ancestry with other primates is but the tip of the iceberg. As we are about to see, the human genome is littered throughout with millions of relics that tell an astonishing story of our evolutionary ancestry.

NOTES

1. I. Sidgwick, "A Grandmother's Tales," *Macmillan's Magazine* 78, no. 468 (October 1898): 433–34, as quoted by J. R. Lucas, "Wilberforce and Huxley: A Legendary Encounter," *Historical Journal* 22 (1979): 314.

2. T. H. Huxley, from a letter dated September 9, 1860, as quoted by J. R. Lucas, "Wilberforce and Huxley: A Legendary Encounter," *Historical Journal* 22 (1979): 326.

3. J. J. Yunis and O. Prakash, "The Origin of Man: A Chromosomal Pictorial Legacy," *Science* 215 (1982): 1525–30.

4. J. W. Ijdo et al., "Origin of Human Chromosome 2: An Ancestral Telomere-Telomere Fusion," *Proceedings of the National Academy of Sciences, USA* 88 (1991): 9051–55.

5. R. Avarello et al., "Evidence for an Ancestral Alphoid Domain on the Long Arm of Human Chromosome 2," *Human Genetics* 89 (1992): 247–49. A. Baldini et al., "An Alphoid Sequence Conserved in All Human and Great Ape Chromosomes: Evidence for Ancient Cen-

tromeric Sequences at Human Chromosomal Regions 2q21 and 9q13," *Human Genetics* 90 (1993): 577–83.

6. The sequences of alphoid segments in region 2q21 (the region where an ancient centromere is located) in human chromosome 2 are in GenBank Accessions AF261193, AF261194, AF261195, AF261198, AF261199, AF261201, AF261202, AF261206, and AF261207 (http://www.ncbi.nlm.nih.gov). I compared the sequences in these accessions with the standard human alphoid sequence using the bl2seq version of the Basic Linear Alignment Search Tool (BLAST) and found thirty-three significant matches to the alphoid sequence in these accessions. The minimum match was 78 percent similar and the maximum 95 percent.

Chapter 2

McCLINTOCK'S MASTERPIECE

Throughout history, rare geniuses have emerged whose works are so extraordinary they deserve to be called masterpieces. In science, the works of Einstein, Pasteur, Newton, Curie, Darwin, and Mendel, among others, come to mind. Any list of twentieth-century scientific masterpieces must include the work of Barbara McClintock. Regrettably, like so many before her, she endured repeated rejection for much of her career before her masterpiece was widely acclaimed.

In 1950, McClintock published what is now a landmark article. She documented a series of detailed experiments and used complicated logic to interpret them.[1] Her experiments and her logic were unassailable, but most scientists initially dismissed her conclusions because they seemed too revolutionary. McClintock later reminisced, "It was a surprise that I was being ridiculed, or being told that I was really mad."[2] A famous biologist reportedly commented at the time, "That woman is either crazy or a genius."[3]

Those are strong words, especially when we consider that the subject of McClintock's revolutionary idea was, of all things, spotted corn kernels. How could corn kernels cause such an uproar? More important, what could they possibly have to do with human evolution? It took almost three decades, but scores of scientists eventually found that McClintock's discoveries in corn also applied to humans. She received the 1983 Nobel Prize in Medicine (a belated thirty-three years after she published her discovery) because her theories for corn advanced our understanding of genetics, evolution, disease, and cancer in humans.

What was so remarkable about spots on corn kernels that led to McClintock's masterpiece? She noticed in the 1940s that the inheritance of these spots didn't quite fit expected patterns, and she set out to discover why. Through years of exhaustive experimentation, coupled with brilliant insight, she surmised that small DNA segments were moving from one place to another in the genome. We now call these mobile pieces of DNA *transposable elements* because they are capable of transposing themselves from one place to another in the genome. Sometimes they are called "jumping genes," but the nickname is a bit misleading.

Transposable elements can disrupt gene function. *Genes* are DNA segments that encode information about an individual's heritable traits. They are translated into protein products that interact and help each new individual grow, survive, and reproduce in ways characteristic of its species. Mutations are the original source of variations in heritable traits. For instance, differences in hair color, eye color, height, and thousands of other characteristics are the result of mutations in the DNA of our ancestors.

Most mutations are substitutions of one base pair for another, such as the following example where a C-G pair is substituted by a T-A pair:

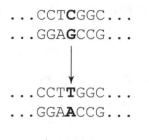

```
...CCTCGGC...
...GGAGCCG...
```

```
...CCTTGGC...
...GGAACCG...
```

substitution

Other common mutations include insertions or deletions of a base pair:

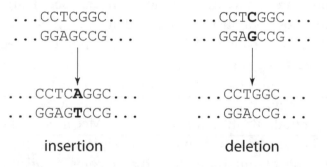

```
...CCTCGGC...          ...CCTCGGC...
...GGAGCCG...          ...GGAGCCG...
```

```
...CCTCAGGC...         ...CCTGGC...
...GGAGTCCG...         ...GGACCG...
```

    insertion   deletion

Once a mutation is established in DNA, it is perpetuated each time the DNA molecule replicates, so a mutation can be passed on from one generation to the next for an untold number of generations.

As McClintock determined, when a transposable element slips into a gene, it causes a drastic mutation that usually eliminates the gene's function:

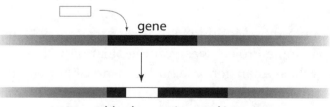

transposable element inserted into a gene

When the element slips out, the gene's original function is often restored:

gene function restored

The liberated transposable element can now insert itself elsewhere in the DNA. If it happens to insert itself in another gene, it causes a new mutation there.

McClintock's discovery that parts of the genetic material could move around in the genome is what caused the uproar. At the time, most scientists thought that once a mutation was established, it remained stable from that point on, replicating itself faithfully from one generation to the next. Moreover, they thought, such mutations had no effect on creating mutations elsewhere; they simply became a stable part of the DNA.

Most mutations fit that picture, and McClintock didn't dispute it. Rather, she saw transposable elements as inducing yet another type of mutation, one that creates *instability* in the genome by moving from one place to another.

McClintock's view was too much for biologists in the 1950s to accept, no matter how valid her experiments were. By the late 1970s and early '80s, however, molecular geneticists had purified and sequenced the DNA of transposable elements from several organisms, including humans. The sequences clearly revealed how the elements move. McClintock gained widespread praise for her work, won a Nobel Prize, and the rest is history.

Well, not quite. McClintock passed away in 1992 just as the Human Genome Project was gaining steam. By then, scientists knew that the human genome had transposable elements—a lot of them. But during the ensuing years as the actual number was gradually revealed, everyone who knew anything about transposable elements was stunned. The human genome contains about *three million* of them, constituting almost half of our DNA!

TRANSPOSABLE ELEMENTS MAY TRANSPOSE OR RETROTRANSPOSE.

There are two types of transposable elements: *transposons* and *retroelements*. McClintock focused on transposons, DNA elements that excise themselves and move to other places in the genome, much like the "cut-and-paste" function of a computer.

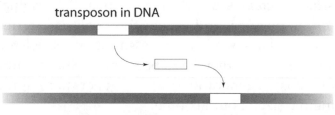

transposon in DNA

transposon moves to a new location in DNA

Transposons make up about 2.8 percent of the human genome, a significant proportion given that genes (the most important part of the genome) make up less than 2 percent. All evidence suggests that human transposons are now completely inactive. About three hundred thousand of them are present in our genome, but they have all mutated so much that they have lost their ability to move. Many generations ago they were actively moving throughout the genome, but now they are stranded in their current positions.

By contrast, *retroelements* make up a whopping 42.8 percent of the human genome. Unlike transposons, retroelements do not excise themselves during transposition. Instead, a retroelement transposes by making an RNA copy of itself, and the RNA copy is then "retrocopied" back into DNA. The DNA copy then inserts itself at a new position. Retroelements use a "copy-and-paste" process instead of the "cut-and-paste" process of transposons.

retroelement in DNA

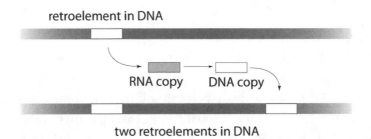

two retroelements in DNA

This process has a profound effect: every time a retroelement transposes, it increases its number by one; the original copy remains in its position, and a new copy of it is inserted elsewhere. This explains why we have so many retroelements in our genome. For ages they have been increasing in number as they transpose, generation after generation, gradually expanding the size of our genome.

Fortunately, of the 2.7 million retroelements currently in our genome, nearly all of them are now inactive, mutated to the point that they can no longer make copies of themselves. The good news is that our genome is now highly stable. The not-so-good news is that it's not entirely stable; a few retroelements are still actively transposing. Scientists have documented brand new cases of retroelement transposition, some of them responsible for genetic disorders and cancer.[4] The overwhelming majority of retroelements, however, are inactive relics of a time in our distant ancestry when transposition was much more active than it is now.

HOW DID TRANSPOSABLE ELEMENTS ORIGINATE?

We cannot fully answer this question for all transposable elements, but for some we can. Many retroelements strongly resemble viruses called *retroviruses*, infectious particles consisting of RNA enclosed in a protective protein coat. They are quite common; among them are the viruses that cause flu and AIDS. Some viruses use DNA as their genetic material, but most viruses that infect humans are retroviruses, which use RNA.

In extremely rare instances in the past, a retrovirus inserted itself into the DNA of an immature reproductive cell that gave rise to sperm or egg cells. After fertilization, repeated cell divisions put the infected DNA into each cell of the new individual. Thereafter, the infected DNA was transmitted from one generation to the next.

Over time, mutations ended the capacity of the integrated viral DNA to escape from the host cell's DNA. No longer could the viral DNA infect new cells. However, it could still replicate along with the DNA in which it resided, and from time to time insert more copies of itself elsewhere in the genome. It had become a retroelement.

All viruslike retroelements in humans originated a very long time ago, tens of millions of years ago in most cases. There is no evidence of modern retroviruses entering the human genome and changing into retroelements. But the same cannot be said for other mammals. Mice, cats, pigs, sheep, chimpanzees, and gorillas all have active retroelements that are relative newcomers. Chimpanzees and gorillas, for example, have multiple copies of a viruslike retroelement, descendants of a retrovirus that independently infected their genomes at about the same time but did not infect the genomes of humans and orangutans.[5]

Other types of retroelements did not originate directly from ancient viruses, although some may have pieces of retroviruses integrated into them. They are opportunistic pieces of genomic DNA that through mutations and shuffling of DNA have acquired the ability to transpose. The most common retroelement in the human genome is a relatively small one called *Alu*, and it is one of these opportunistic retroelements. About 10 percent of the human genome consists of more than one million *Alu* elements. *Alu* is also one of the few types of retroelements still actively transposing in the human genome.

WHAT DO TRANSPOSABLE ELEMENTS
TELL US ABOUT HUMAN EVOLUTION?

Even more than transposons, retroelements offer intriguing glimpses into our evolutionary past. They can insert themselves anywhere in our DNA and tend to remain stuck in place. The chance of two retroelements independently inserting themselves into exactly the same position in the genome in two different individuals is exceptionally small. It follows that if two individuals have a retroelement at exactly the same location, they must have inherited that retroelement from a common ancestor.

What do we find when we compare the positions of retroelements in humans and other primates? Are there transposable elements in the same places? The first evidence emerged almost by accident in the mid-1980s. When scientists purify the DNA of a gene, some of the DNA surrounding the gene usually comes along with it. Gene after gene, scientists found transposable elements, especially retroelements, in the DNA surrounding these human genes, and in some cases within the genes themselves. These elements, it seemed, were everywhere. Soon, scientists isolated the same genes and their surrounding DNA from other primates and compared the DNA sequences with those isolated from humans.

Through the course of many studies an unmistakable pattern emerged. In case after case, transposable elements in human DNA were present at exactly the same positions in chimpanzee DNA, and to a lesser degree in other apes and monkeys. The human and chimpanzee versions of these elements were highly similar, about 98 percent similar, but not identical. These observations could best be explained if the transposable elements became established in the DNA of a common ancestor of humans and chimpanzees, then they mutated in the separate lineages.

Consider three examples. The first is a study of the DNA surrounding a set of genes that encode hemoglobin, a protein that carries oxygen and gives blood its red color. In 1985, scientists at the University of California, Davis, and the University of California, Berkeley, compared human and chimpanzee DNA from this region and found several *Alu* elements.[6] As it turns out, all of the *Alu* elements are in exactly the same places, and same orientations, in both species.

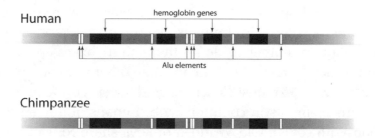

The researchers compared the DNA sequences of seven of the human and chimpanzee *Alu* elements and found that they are highly similar but not identical, ranging from 94.7 percent to 98.9 percent similarity. Three to fifteen mutations in each element account for the differences. Common ancestry is the only reasonable explanation of such strong similarity in the sequences and positions of these elements.

The second example is HERV-K, a viruslike retroelement that entered the genome of a common ancestor of humans, apes, and monkeys tens of millions of years ago. Unlike most viruslike retroelements, it is still actively transposing in the human genome.

Scientists at the Shemyakin-Ovchinnikov Institute of Bioorganic Chemistry in Russia carried out an ingenious study to determine when the different copies of HERV-K elements were inserted into the genome during primate evolution.[7] They began with fourteen known HERV-K elements in the human genome then determined the DNA sequences flanking both sides of each element. They then searched for the same flanking sequences in chimpanzees, gorillas, orangutans, gibbons, Old World monkeys, and New World monkeys. The presence or absence of HERV-K at comparable places in the genomes would be evidence of how closely related different species are to humans.

Three of the fourteen HERV-K elements were present only in the human genome, so they were inserted after the divergence of the lineages leading to humans and the next most closely related species. The remaining eleven were present in the same positions in humans, chimpanzees, and gorillas, which reinforces other evidence that chimpanzees and gorillas are more closely related to humans than other apes or any of the monkeys. The next most closely related species is the orangutan with nine elements in common with humans, followed by the gibbon with seven. Old World monkeys share four of the elements with humans, whereas the New World monkeys share only two.

We can summarize the relationships in an evolutionary tree diagram:

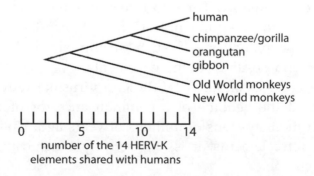

This "family tree" reflects only the Russian study, independent of the traditional relationships of humans, apes, and monkeys. Even so, it is consistent with our traditional understanding of relationships derived from the fossil record, geography, and many other DNA studies. The strong convergent evidence shows that humans are most closely related to the African apes (chimpanzees and gorillas), followed by the Asian apes (orang-utans and gibbons), with the monkeys more distantly related.

Our third example highlights CMT1A, a duplicated segment of DNA with *Alu* elements on both ends. Humans and chimpanzees have two copies of the CMT1A segment at exactly the same places in their genomes. However, the gorilla and orangutan genomes each have only one copy of the CMT1A segment, in the same position as one of the CMT1A segments in humans and chimpanzees. The two copies of CMT1A in humans and chimpanzees are similar but not identical. In particular, an *Alu* element on the end of one of the copies is truncated (missing part of its DNA) in both humans and chimpanzees when compared with its corresponding *Alu* element in the other copy.

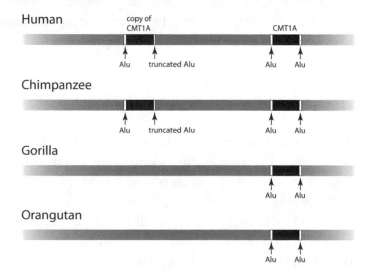

The presence of the duplicated segment in humans and chimpanzees, and its absence in gorillas and orangutans, suggests that the duplication happened in the common ancestor of humans and chimpanzees, most probably after that ancestor had separated from the ancestor of gorillas and orangutans. It implies that humans and chimpanzees are more closely related to each other than either is to gorillas or orangutans.

It's no coincidence that the duplicated CMT1A segment has retroelements on its ends. It is now well known that repeated sequences (such as retroelements) located close to each other target the DNA between them for duplication, which explains the duplication of CMT1A in humans and chimpanzees. The newly duplicated segments now predispose any DNA between them to further duplication. Duplication of the sequence between the two CMT1A adds another copy of a gene called *PMP22* that lies between the two CMT1A segments. The two copies of this gene cause Charcot-Marie tooth disease, which has serious neurological consequences. In fact, the CMT1A segment was named after the disease ("CMT" in CMT1A stands for Charcot-Marie tooth disease). Because of the tendency of the segment containing *PMP22* to further duplicate, this disease is relatively common, affecting about one in twenty-five hundred people, an unfortunate consequence of our evolutionary heritage.

A GENETICALLY ENGINEERED KISS AWAKENS SLEEPING BEAUTY.

Sleeping Beauty, a transposable element that in a strange way resembles its fairy-tale namesake, is one of the most captivating stories in recent science. To understand the story, we need to return to transposons, the DNA elements first discovered by Barbara McClintock that transpose through a cut-and-paste process. Recall that in humans, all transposons are highly mutated and inactive—asleep in a sense.

A particular type of transposon, called Tc-1/mariner, is widespread in nature, found in insects, mammals (including humans), reptiles, amphibians, and fish, and is probably present in all vertebrates. It is also inactive in all vertebrates, mutated to the point that it can no longer transpose—and it has been that way for millions of years. In the mid-1990s, a group of scientists at the University of Minnesota had a brilliant idea. With so many copies of this element (about fourteen thousand in the human genome alone) and so many species with it, could they compare the DNA sequences of the many copies and extrapolate back to find the original sequence? If so, could they use genetic engineering to reconstruct the original element and see if it could once again transpose?

They chose to work with Tc-1/mariner elements from salmon and trout because these elements had acquired the fewest mutations and would require the least tinkering to restore the original DNA sequence. They then compared the sequences of these elements in different species of trout and salmon and used a "majority-rule" approach to determine the original sequence; wherever there was a difference in the DNA sequences, they assumed that the sequence in the majority of elements was the original. Ultimately, they found what is apparently the original sequence and used genetic engineering to re-create the element.

When they injected this engineered element into cultured fish cells, it transposed itself into the DNA of the cells. It did the same thing in cultured cells of other species, including human cells. By extrapolating the evolutionary history of a transposon, they had awakened it after it had lain asleep for millions of years. Appropriately, they named it Sleeping Beauty.[8]

The awakening of Sleeping Beauty confirms that reversing mutations can restore the function of inactive transposable elements. As remarkable as this discovery is for academic science, it also has great promise in medicine, giving hope to tens of thousands of people who suffer from debilitating genetic disorders like sickle-cell anemia and cystic fibrosis. People who have these disorders carry mutated versions of genes. If, somehow, a corrected version of the gene could be delivered to the proper cells, and if that gene could then be inserted into the DNA of those cells, these people could potentially live their lives free of the horrific suffering caused by these disorders.

That hope is now reality for a few genetic disorders through a procedure called gene therapy. Surgeons extract cells from an affected patient, then take a genetically engineered retrovirus that carries a corrected version of the mutated gene in it and induce that virus to infect the cells. In some cells, the virus delivers a DNA copy of itself, along with the corrected gene, into the DNA of the cells. Then, surgeons return the cells to the patient where, if all goes well, the corrected gene functions normally.

One of the problems with gene therapy, however, is the low rate at which genetically engineered viruses insert DNA copies of themselves into cellular DNA. Scientists who work with fruit flies found that active transposons were a much better tool for inserting corrected genes into cells, and there are plenty of active transposons in fruit flies, but none in humans—until Sleeping Beauty was awakened. Research is now under way to test Sleeping Beauty and other genetically engineered transposons, such as the Frog Prince, a restored transposon from frogs, as tools for delivering corrected genes to people who need them.[9]

ARE TRANSPOSABLE ELEMENTS USEFUL?

Genetically engineered transposons will probably prove useful for medicine. But what about the millions of transposable elements that make up almost half of our genome? Do they naturally offer some useful functions? So far, it appears that nearly all of them reside in places where they do neither harm nor good. Apparently they are just along for the ride, and most have been riding for a very long time. It is tempting to think of them as millions of ancient but now mostly benign parasites trapped within our genome. Some scientists have even called them "junk DNA," implying that they are useless and, currently, harmless relics of evolution.

The majority probably are benign. However, a few retroelements are still transposing and they can be dangerous when they promote genetic disorders or cancer. Interestingly, a few transposable elements have taken on productive functions. By chance, some inserted themselves near genes where they influence how those genes operate. For this reason, Barbara McClintock called them "controlling elements" because those located near genes may reprogram the activity of those genes. Surprisingly, even a few human genes are composed almost entirely of transposable elements, suggesting that these elements combined with one another and mutated to create DNA sequences that now function as useful genes.[10] These, however, are just a tiny fraction of the transposable elements that make up almost half of our DNA. Most of them appear to be completely useless.

Regardless of their function, or more often the lack of it, transposable elements provide powerful evidence of human evolution and our common ancestry with chimpanzees and other primates. They are by far the most abundant relics of our evolutionary history. Beyond being relics themselves, transposable elements play a role in adding to the genome yet another class of relics that masquerade as genes. These bogus genes littering our genome are the theme of the next chapter.

NOTES

1. B. McClintock, "The Origin and Behavior of Mutable Loci in Maize," *Proceedings of the National Academy of Sciences, USA* 36 (1950): 344–55.

2. E. F. Keller, *A Feeling for the Organism: The Life and Work of Barbara McClintock* (San Francisco, CA: W. H. Freeman, 1983), p. 140.

3. Ibid., p. 142.

4. P. L. Deininger and M. A. Batzer, "*Alu* Repeats and Human Disease," *Molecular Genetics and Metabolism* 67 (1999): 183–93.

5. C. T. Yohn et al., "Lineage-Specific Expansions of Retroviral Insertions within the Genomes of African Great Apes but Not Humans and Orangutans," *PLoS Biology* 3, no. 4 (2005): e110, http://biology.plosjournals.org.

6. I. Sawada et al., "Evolution of *Alu* Family Repeats since the Divergence of Human and Chimpanzee," *Journal of Molecular Evolution* 22 (1985): 316–22.

7. Y. B. Lebedev et al., "Differences in HERV-K LTR Insertions in Orthologous Loci of Humans and Great Apes," *Gene* 247 (2000): 265–77.

8. Z. Ivics et al., "Molecular Reconstruction of Sleeping Beauty, a Tc1-like Transposon from Fish, and Its Transposition in Human Cells," *Cell* 91 (1997): 501–10.

9. C. Miskey et al., "The Frog Prince: A Reconstructed Transposon from *Rana pipiens* with High Transpositional Activity in Vertebrate Cells," *Nucleic Acids Research* 31 (2003): 6873–81.

10. R. J. Britten, "Coding Sequences of Functioning Human Genes Derived Entirely from Mobile Element Sequences," *Proceedings of the National Academy of Sciences, USA* 101 (2004): 16825–30.

Chapter 3

BOGUS GENES

One of the most valued organisms in the world is the microscopic single-celled fungus known as baker's or brewer's yeast. It ferments bread dough to make it rise and it ferments beer during the brewing process. It also was one of the first organisms to have all the DNA in its genome sequenced. At thirteen million base pairs, its genome is relatively small; a human cell has almost two hundred and fifty times as much DNA.

From an evolutionary perspective, we (along with all other animals and plants) trace our ancestry to single-celled microscopic organisms with a cell nucleus, like yeast. If that is so, we ought to find genes in the yeast genome that resemble our own genes—and we do. However, we have many more genes than yeast does. So, if we, along with all animals and plants, have ancient single-celled ancestry, how did we get all of those extra genes?

The DNA sequences of our genes, and the genes of many other organisms, hold powerful clues. Most of our genes resemble other genes in our genome. We can best explain this situation if we assume that new genes hardly ever arise from scratch. Instead, they are derived from duplication of existing genes. Once a gene has been duplicated, only one copy is needed to carry out the gene's original function, so the other copy is free to mutate. After enough mutations have accumulated, this second copy may end up as a new gene with a new function.

Gene duplication followed by mutational divergence of the copies is how the number of genes in a genome increases over many generations during evolution. Gene duplication is a rare event, but new duplications are well documented. The geneticist Alfred Sturtevant was the first to discover new duplications. While working with fruit flies in 1925, he discovered that a few offspring had an extra copy of a certain gene when compared to their parents.[1] Other scientists have since observed new gene duplications in many different species, including humans.

If duplication followed by divergence is indeed the mechanism for increasing gene number, how do cells "know" which mutations to make in a duplicated gene so that it works? If mutations are random events, many of them should disable the mutated gene, rendering it useless. As abundant evidence from genome studies suggests, *this is precisely what happens*. Quite a few of our genes were duplicated not once but many times, and some of the copies mutated into useless relics of genes. These leftover, nonfunctional copies of genes are known as *pseudogenes*.

THERE ARE THREE MAJOR TYPES OF PSEUDOGENES.

Pseudogenes can arise by one of three processes, each of which leaves its own molecular signature. We classify pseudogenes into three types based on those processes: unitary pseudogenes, duplication pseudogenes, and retropseudogenes.

The first type, a unitary pseudogene, is a mutated form that has no functional copy of itself anywhere else in the genome. It functioned at one time but doesn't now. Most genes carry out important roles. But as environments change over time and as organisms evolve to exploit new environments, a gene that at one time was essential may no longer be needed. At that point, mutations that disable the gene are no longer detrimental to the organism, and the mutated, nonfunctional gene can persist with no ill effects. It is now a pseudogene at the same position as the original gene. If there are no other copies of the gene in the genome, it is a unitary pseudogene.

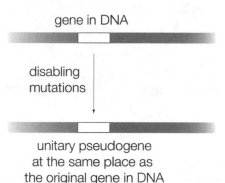

gene in DNA

disabling mutations

unitary pseudogene
at the same place as
the original gene in DNA

Unitary pseudogenes often correspond to vestigial organs, which lost their function when they were no longer needed. Moles, for example, spend nearly all of their lives underground in complete darkness. Although they have vestiges of eyes, some species of moles have completely lost the ability to see because vision is of no benefit to them in the absolute darkness of their burrows. The marsupial mole, a relative of kangaroos and possums, is one of the blind species. It is native to the sandy deserts of Australia, where it spends nearly all of its life burrowing under the sand, only very rarely venturing above ground and then usually in the darkness of night after a rain. It has remnants of eyes, but they are completely useless; the nerve that connects eyes to the brain is missing. In 1997, a group of scientists in the United States and Northern Ireland found that one of the essential genes for vision in mammals is highly mutated in marsupial moles.[2] There are no other copies of the gene in the mole's genome, so the function of that gene is completely lost. The gene is now a unitary pseudogene.

Duplication pseudogenes are the second type. (They are also called redundant pseudogenes or unprocessed pseudogenes.) Occasionally, a segment of DNA containing a gene is duplicated to produce two copies that end up next to each other. If one mutates into a nonfunctional copy, it becomes a duplication pseudogene.

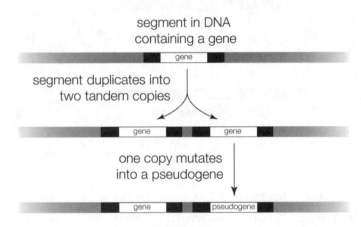

Existing duplications are prone to further duplication, resulting in three, four, or more copies of the same segment, all in tandem with one another. Approximately 3,300 human genes (about 15 percent of our genes) reside in clusters of highly similar genes, suggesting that these clusters arose through multiple rounds of tandem duplication.[3] When scientists examined copies of genes within the clusters, they found that accumulated mutations ultimately have one of two possible effects. Some mutations produce an altered but functional copy that carries out a new function. Other mutations disable the gene, and it becomes a duplication pseudogene. We have about ten thousand duplication pseudogenes in our genome, many within the tandem clusters of duplicated genes.

The third type is called a retropseudogene (also a processed pseudogene). It originates in almost the same way as retroelements. A gene typically produces multiple copies of RNA copied from its DNA as part of its normal function. Occasionally, the products made by retroelements mistakenly recognize the RNA copied from a gene as if it were a retroelement. They make a DNA copy of it, which then inserts itself elsewhere in the DNA as if it were a retroelement.

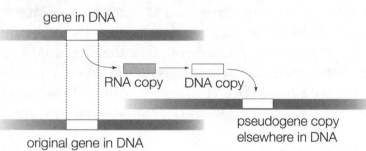

Even if it is not mutated, this new copy of the gene usually does not function. For a gene to function, it needs to have certain important DNA sequences nearby to regulate it. When the new copy is inserted elsewhere, it is no longer near its regulatory sequences and it cannot function. It, too, is a pseudogene.

Certain genes have multiple retropseudogene copies of themselves in the genome. For example, one of our most important genes is the cytochrome c gene, which encodes one of our most ancient proteins, one that allows us to thrive in an oxygen-rich world. Scattered throughout our genome are forty-nine nonfunctional copies of the gene, most of them retropseudogenes.[4]

HOW MANY PSEUDOGENES DO WE HAVE?

Identifying and counting the number of pseudogenes in the human genome is no easy task. As DNA-sequencing machines churned out the sequence information of the human genome, a large group of scientists used specialized DNA analysis software to look for genes in the assembled sequence. One of their greatest challenges was distinguishing true genes from pseudogenes. When the first draft of the human genome was published in 2001, genome scientists estimated that the number of true genes was around thirty-five thousand. They had already identified and excluded many pseudogenes, but the thirty-five-thousand-gene estimate turned out to be too high. With exhaustive and detailed experimental analysis, they weeded out even more pseudogenes. In October of 2004, they reported that the number of confirmed true genes is 19,438 and the number of predicted true genes (DNA sequences with all the characteristics of true genes and no identifiable mutations to make them pseudogenes) is 2,188.

In the meantime, scientists were also tallying the number of confirmed pseudogenes, but the work was tricky. Many transposable elements contain pseudogenes, some resembling genes found in viruses. If those were added in, the total number would be in the millions. However, an informative count of pseudogenes should exclude any sequences from transposable elements, targeting only those pseudogenes that closely resemble true human genes. Even then, the number is still substantial. By late 2003, the count stood at 19,724, slightly more than the number of true genes.[5]

The researchers who compiled this number admitted that the methods they used were conservative to avoid the possibility of misidentifying a pseudogene. Pseudogenes often elude detection because they may be fragmented, with pieces scattered throughout the genome. Older pseudogenes often accumulate so many mutations that they lose much of their resemblance to the original gene, so computer programs designed to search for them in DNA sequence databases may miss them. In the words of the study's authors, "It is highly likely that the human pseudogenes identified here represent only a small fraction of the total."[6] In other words, the number of pseudogenes in our genome exceeds, and probably greatly exceeds, the number of real genes.

WHAT DO PSEUDOGENES TELL US ABOUT HUMAN EVOLUTION?

Let's look at just a few examples of pseudogene evolution in humans, starting with a unitary pseudogene called the *GULO* pseudogene. There is no functioning *GULO* gene in humans, but the functional version is present in many other animals, where it allows them to make vitamin C. Animals with a functional *GULO* gene do not need to consume vitamin C in their diets because their cells can make it from other substances. Dogs and cats, for example, eat foods that contain hardly any vitamin C but their *GULO* gene allows their bodies to produce their own vitamin C. By contrast, humans and all other primates lack this gene and must consume vitamin C to survive.

So why don't humans have a *GULO* gene? Actually, a piece of the gene is still in our DNA, but it is now a highly mutated unitary pseudogene with much of the original gene missing. The fragment that's left is loaded with mutations; about 20 percent of the DNA sequence is mutated when compared to the functioning version in other mammals.[7] To accumulate so many mutations, the original gene must have been disabled long ago in our evolutionary ancestry. All primates have the same highly mutated fragment so the gene probably was disabled before different primate lineages diverged from their common ancestor.

Although this pseudogene is highly mutated and utterly useless, humans and chimpanzees have almost identical copies of it. The chimpanzee genome contains the same *GULO* pseudogene, in the same place in the DNA, and the chimpanzee and human versions are 98 percent identical.[8] The same can be said for almost every other pseudogene in the human genome. With a few notable exceptions, chimpanzees and humans have the same pseudogenes in the same places, and they are, on average, about 98 percent similar.

How could such a beneficial gene be present in some species but end up as a pseudogene in others? Darwinian natural selection gives us a perfectly reasonable explanation. Species whose diets do not contain much vitamin C, such as carnivores, require a functional *GULO* gene. For these species, individuals who have only a mutant version of this gene die of vitamin C deficiency, ultimately eliminating the mutated gene.

However, the diets of primates (including humans) typically contain adequate vitamin C because they include fruits and vegetables. Fossilized teeth and other evidence indicate that ancient primates also consumed plants with adequate vitamin C, so there was no need for their bodies to produce it. Disabling mutations in the *GULO* gene had no disadvantage because the high amount of vitamin C in the diet rendered the gene unnecessary. Millions of years later when humans began to venture out to sea for long periods of time, the fruits and vegetables they carried with them spoiled so quickly that their diets at sea became vitamin C deficient, resulting in scurvy. The need for a functional *GULO* gene had returned, but all that remained of it was a useless and highly mutated unitary pseudogene.

Let's now turn our attention to duplication pseudogenes, starting with a gene whose name is a mouthful: the glucocerebrosidase gene, which we'll call by its abbreviation, *GBA*. The human genome has one functional *GBA* gene and a tandem pseudogene copy nearby. The pseudogene copy fails to function because a segment of fifty-five base pairs is missing. In 2005, scientists at the University of Victoria in Canada found that chimpanzees and gorillas have the same duplication pseudogene in the same position and with the same fifty-five base-pair deletion as in humans.[9] Orangutans, on the other hand, have the same two copies of the *GBA* gene in the same places as humans, chimpanzees, and gorillas, but both copies of the gene function; the fifty-five base pairs are there. Squirrel monkeys, by contrast, have only a single copy of the *GBA* gene.

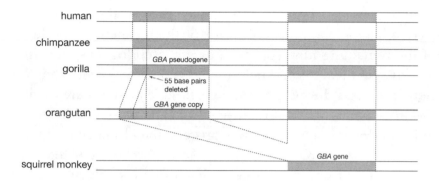

The fifty-five base-pair deletion in the *GBA* pseudogene is an especially telling mutation because it is very unlikely that it would ever originate more than once. A single base-pair mutation at a particular place is rare but can happen independently several times and thus may not always be attributable to common ancestry. However, a deletion of fifty-five base pairs in exactly the same place in the same pseudogene in different species is strong evidence of a common ancestor.

Most likely the lineage leading to monkeys diverged from the one leading to humans and great apes when only one copy of the *GBA* gene existed. The gene then became duplicated in the common ancestor of humans and great apes. Later, one of the copies suffered a fifty-five base-pair deletion in the lineage leading to humans, chimpanzees, and gorillas after it had diverged from the lineage leading to orangutans. According to this scenario, humans, chimpanzees, and gorillas are more closely related to one another than any of the three is to orangutans. By the same reasoning, the fact that humans and all of the great apes (including orangutans) share the same duplication of the *GBA* gene implies that humans and great apes are more closely related to one another than they are to monkeys.

Let's now look at one of our oldest and best-studied duplication pseudogenes. As mentioned in the previous chapter, hemoglobin is a vital oxygen transporter in the blood. The human genome contains thirteen copies of the genes that encode hemoglobin, but only four of them function in adults. Of the remaining nine copies, four are pseudogenes, and five are active in the fetus and help the fetus obtain oxygen from the mother's blood during pregnancy. These fetal genes are turned off either before or shortly after birth. The hemoglobin genes and pseudogenes reside in two clusters, the alpha cluster with four genes and three pseudogenes, and the beta cluster with five genes and one pseudogene. Each of these clusters arose anciently from repeated tandem duplication of an original gene.

The single pseudogene in the beta cluster is called the psi-beta pseudogene and it is full of mutations; about 30 percent of its DNA sequence is mutated when compared to its parent gene. To accumulate so many mutations, this pseudogene must have arisen through duplication and then been disabled a very long time ago.

If we compare the beta cluster in six primate species, a fascinating pattern of evolution emerges. Figure 3.1 shows the alignment of genes and pseudogenes in the beta cluster of humans, chimpanzees, gorillas, baboons, New World (American) monkeys, and lemurs (small, rodent-sized primates that are neither apes nor monkeys). Humans, chimpanzees, and gorillas have the same copies of the genes and the psi-beta pseudogene in the same positions, so these three species must be closely related. Baboons have the same copies in the same positions but the delta gene mutated into a new pseudogene. New World monkeys have the psi-beta pseudogene, but humans, chimpanzee, gorillas, and baboons have one more hemoglobin gene than they do. The extra gene arose when the gamma gene duplicated into two copies in their common ancestor. In the lineage that led to lemurs, a piece of DNA was deleted between the psi-beta pseudogene and the delta gene. The deletion disabled the delta gene and fused a piece of it with a piece of the psi-beta pseudogene. The outcome was a new pseudogene called the psi-beta-delta pseudogene.

The presence of the psi-beta pseudogene, or a piece of it, in all primates reinforces the meaning of its large number of mutations: it became a pseudogene long ago in a common ancestor of all primates, whereas other pseudogenes arose more recently.[10]

Let's now turn to another duplication pseudogene in humans, the psi-zeta pseudogene, which lies in the alpha cluster. This pseudogene is almost identical with the nearby zeta gene, a functioning fetal gene. However, a single mutation present in all humans completely disabled it. Because it is so similar to a

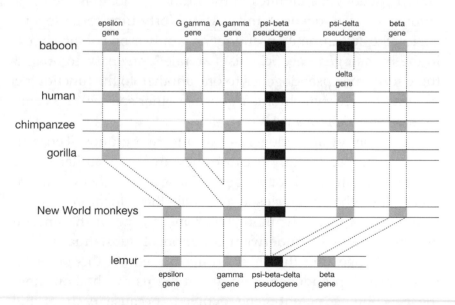

Figure 3.1. The beta cluster of hemoglobin genes in baboons, humans, chimpanzees, gorillas, New World monkeys, and lemurs. True genes are colored gray and pseudogenes are colored black.

true gene, it must have arisen recently; insufficient evolutionary time has passed for a large number of mutations to accumulate in it. The mutation that disabled the second copy of the zeta gene is present only in humans. Chimpanzees have two functional copies of the zeta gene, one in the same place as the human zeta gene and the other in the same place as the human psi-zeta pseudogene. The human version must have mutated into a pseudogene *after* the ancestral divergence of humans and chimpanzees.[11]

Retropseudogenes constitute the third type of pseudogene, and our genome and the genomes of other primates have an unusually large number of them. The most likely explanation is that primates have an enormous number of retroelements, which encode the products that make retropseudogenes.

In some cases, multiple retropseudogene copies of a single gene collectively represent a pseudogene family. One of the largest pseudogene families in the human genome is the cytochrome *c* pseudogene family. With forty-nine pseudogene copies, the cytochrome *c* pseudogene family is one of the largest in the human genome. Scientists at Yale University compared these forty-nine pseudogenes to one another, to the functioning gene, and to the cytochrome *c* genes in other species, and came to some surprising conclusions.[12]

First, the retropseudogenes fall into two groups. Those in one group arose recently because they are quite similar to the modern human cytochrome *c* gene. Those in the other group are much older; they more closely resemble the cytochrome *c* gene in rodents. Apparently, the cytochrome *c* gene in the lineage leading to humans underwent a period of substantial evolutionary change after the older group formed. As this observation shows, a pseudogene family can tell us much about how the functioning parent gene evolved, because each of the pseudogenes leaves a record of the parent gene's sequence at the time it was formed.

Among the human cytochrome c pseudogenes is a strange one. Rodents have a second copy of the cytochrome c gene that functions only in the testes. It turns out that humans also have a copy of this testes-specific gene, but it is a nonfunctional pseudogene that differs from all of the others. It probably lost its function long ago.

When scientists compare pseudogene sequences among species, the same pattern emerges over and over. Human pseudogenes are most similar to those in chimpanzee DNA and, to a lesser extent, are highly similar to those of other primates. Species as distant as rodents and humans also show some similarity in ancient pseudogenes. Once again, we find evidence of our shared evolutionary ancestry with other primates, and more distant shared ancestry with other mammals. But *how* closely are we related to other primates? That is the topic of the next chapter.

NOTES

1. A. H. Sturtevant, "The Effects of Unequal Crossing over at the *Bar* Locus in *Drosophila*," *Genetics* 10 (1925): 117–47.

2. M. S. Springer et al., "The Interphotoreceptor Retinoid Binding Protein Gene in Therian Mammals: Implications for Higher Level Relationships and Evidence for Loss of Function in the Marsupial Mole," *Proceedings of the National Academy of Sciences, USA* 94 (1997): 13749–59.

3. International Human Genome Sequencing Consortium, "Finishing the Euchromatic Sequence of the Human Genome," *Nature* 431 (2004): 931–45.

4. Z. Zhang and M. Gerstein, "The Human Genome Has 49 Cytochrome c Pseudogenes, Including a Relic of a Primordial Gene that Still Functions in Mouse," *Gene* 12 (2003): 61–72.

5. D. Torrents et al., "A Genome-wide Survey of Human Pseudogenes," *Genome Research* 13 (2003): 2559–67.

6. Ibid.

7. Y. Inai, Y. Ohta, and M. Nishikimi, "The Whole Structure of the Human Nonfunctional L-gulono-gamma-lactone Oxidase Gene—the Gene Responsible for Scurvy—and the Evolution of Repetitive Sequences Thereon," *Journal of Nutritional Science and Vitaminology* 49 (2003): 315–19; Y. Inai and M. Nishikimi, "Random Nucleotide Substitutions in Primate Nonfunctional Gene for L-gulono-gamma-lactone Oxidase, the Missing Enzyme in L-ascorbic Acid Biosynthesis," *Biochimica et Biophyisica Acta* 1472 (1999): 408–11.

8. The sequence similarity cited here is based on a computer-based (BLAST) search of the chimpanzee genome, which I conducted on February 7, 2005, with the published sequence of the human *GULO* pseudogene as the query sequence. The sequence similarity is 554 identities of a total of 565 base pairs consisting of ten substitutions and one deletion in the chimpanzee sequence when compared to the human sequence. The two sequences are located at the same positions on chromosome 8 in the human and chimpanzee genomes.

9. J. R. Wafaei and F. Y. Choi, "Glucocerebrosidase Recombinant Allele: Molecular Evolution of the Glucocerebrosidase Gene and Pseudogene in Primates," *Blood Cells, Molecules, and Diseases* 35 (2005): 277–85.

10. S. Harris et al., "The Primate $\psi\beta1$ Gene: An Ancient b-globin Pseudogene," *Journal of Molecular Biology* 180 (1984): 785–801; L.-Y. E. Chang and J. L. Slightom, "Isolation and Nucleotide Sequence Analysis of the β-type Globin Pseudogene from Human, Gorilla, and Chimpanzee," *Journal of Molecular Biology* 180 (1984): 767–84.

11. W. C. Wong et al., "Comparison of Human and Chimpanzee Zeta 1 Globin Genes," *Journal of Molecular Evolution* 22 (1985): 309–15; N. J. Proudfoot, A. Gil, and T. Maniatis, "The Structure of the Human Zeta-Globin Gene and a Closely Linked, Nearly Identical Pseudogene," *Cell* 31 (1982): 553–63.

12. Zhang and Gerstein, "The Human Genome Has 49 Cytochrome *c* Pseudogenes."

Chapter 4

SOLVING THE TRICHOTOMY

For most of the twentieth century, scientists classified chimpanzees, gorillas, orangutans, and gibbons as the Pongidae, a single family of apes. They kept humans separate, as the only modern species of the family Hominidae, which includes *Homo neanderthalensis* and other extinct humanlike species. Do DNA studies confirm this separation?

The answer is a resounding no. As scientists finally realized, the human, chimpanzee, and gorilla branches of the primate family tree are closer to one another than they are to the orangutan and gibbon branches. But it took some time to solve the *trichotomy problem*: of humans, chimpanzees, and gorillas, which two of the three are most closely related to each other? Until large-scale DNA studies were available, it didn't have a clear answer. To understand how scientists found the answer, we need to jump back through eons of time to explore the origins of an odd little genome found in all animals and plants: the tiny but essential genome in mitochondria.

MITOCHONDRIA LOOK LIKE BACTERIA TRAPPED IN OUR CELLS.

So far we've focused our discussion on the DNA in chromosomes, which are located in the nucleus of each cell; the nucleus is where nearly all of our DNA resides. However, there is a small amount of DNA in a set of microscopic, bean-shaped structures within our cells called *mitochondria* (singular *mitochondrion*). The DNA in human mitochondria has a mere thirteen genes, compared to more than twenty thousand genes in the chromosomal DNA. But those thirteen genes are absolutely essential for our survival.

By the 1960s, scientists had found multiple lines of evidence showing that mitochondria are very similar to bacteria. For example, as our cells grow, the mitochondria inside of them reproduce, dividing in two by a mechanism that resembles bacterial cell division. The DNA molecules in both mitochondria and bacteria are circular, unlike the linear DNA in chromo-

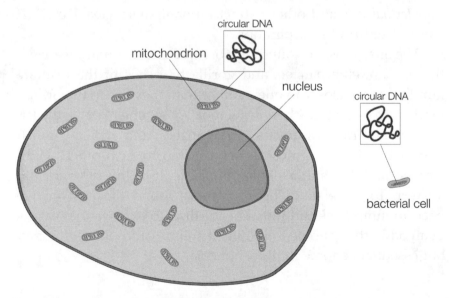

Figure 4.1. Comparison of a human cell and a bacterial cell. Like bacterial cells, mitochondria contain circular DNA.

somes (figure 4.1). Moreover, mitochondrial genes are more like bacterial genes than chromosomal genes. There are many other similarities between mitochondria and bacteria, far too many for them to be coincidental.

Chloroplasts in plant cells also resemble bacteria. These microscopic football-shaped structures give leaves their green color and carry out photosynthesis, the process that traps energy from sunlight. Like mitochondria, chloroplasts contain circular DNA, their genes resemble bacterial genes, and they grow and divide like bacteria.

Recognizing the similarities that bacteria share with mitochondria and chloroplasts, Lynn Margulis, currently a professor at the University of Massachusetts, postulated that these structures trace their evolutionary origins to bacteria and that they evolved through a symbiotic relationship with the cells in which they reside, an idea called the *endosymbiotic hypothesis*. To understand this hypothesis, we need to look deep into the history of life on Earth. In rocks formed from sediments deposited on the ocean floor about 570 million years ago, we find what at one time people thought were the oldest fossils. The emergence of these fossils, strange sea-dwelling organisms including the trilobites, is known as the *Cambrian radiation* (sometimes called the Cambrian explosion). The apparent absence of fossils in rocks older than the Cambrian period led some people to speculate that the Cambrian radiation was the beginning of creation. Some opponents of evolution continue to use this page in the fossil record as evidence of a sudden creation.

However, when scientists carefully studied rocks that predated the Cambrian radiation, they discovered that the oldest fossils are actually much, much older than Cambrian fossils. Nearly all of these older fossils are microscopic, which is why it took so long for people to find them. Some evidence suggests that microscopic life may date back as far as 3.5 billion years ago, although the earliest undisputed fossils are about 2.5 billion years old. All of the oldest fossils are of bacteria, which do not have a cell nucleus. They are classified as *prokaryotes*, which means "before nucleus." Organisms whose cells do have a nucleus (like humans, animals, plants, and fungi) are classified as *eukaryotes*, which literally means "true nucleus." The earliest fossils with nucleated cells are from microscopic eukaryotes, dating to about 1.8 billion years ago.

According to the endosymbiotic hypothesis, bacteria infected ancient eukaryotic cells and ended up becoming an essential part of those cells long before any multicellular organisms appeared. The engulfed bacteria became dependent on their host cells for some of their needs, and the eukaryotic host cells became dependent on the bacteria inside of them to carry out essential functions. The relationship was *symbiotic*, a mutually beneficial situation. Hence, the word *endosymbiosis*, *endo*-meaning "within" to denote the engulfed bacterial cells that resided within ancient eukaryotic cells.

There were at least two major endosymbiotic episodes: the first produced mitochondria and the second chloroplasts. The lineage that produced animals arose directly from the first episode. The second episode happened in cells that already had primitive mitochondria and introduced a second type of endosymbiotic bacteria that became chloroplasts. These cells were the ancestors of plants. The divergences that led to animals and plants happened while all organisms on Earth were still microscopic.

DOES EVIDENCE FROM DNA AND GENOME ANALYSIS SUPPORT THE ENDOSYMBIOTIC HYPOTHESIS?

When evidence favoring the endosymbiotic hypothesis was first presented, methods for DNA analysis, which could either make or break the hypothesis, were still being developed. If the hypothesis were correct, then DNA sequences in mitochondria and chloroplasts should most closely resemble those in bacteria. Also, it should be possible to identify the closest living bacterial relatives of mitochondria and chloroplasts by comparing bacterial, mitochondrial, and chloroplast genomes.

It took decades for the DNA evidence to accumulate, but little by little it consistently confirmed the endosymbiotic hypothesis. Some of the first solid evidence showed a connection between chloroplasts and a type of bacteria called cyanobacteria. Both cyanobacteria and chloroplasts carry out photosynthesis and share many of the same genes. In many respects, chloroplasts look like cyanobacteria trapped in plant cells. Cyanobacteria apparently are the closest living relatives of chloroplasts in modern plants.

DNA evidence about mitochondria accumulated more slowly. Bacterial genomes were fully sequenced in the late 1990s and early 2000s. When they were compared with mitochondrial genomes, the *Rickettsiae* emerged as the closest relatives of mitochondria. Modern bacterial species within this group live, grow, and divide naturally within eukaryotic cells, much like mitochondria do. Unlike mitochondria, they can briefly escape from their host cells and infect other cells. They often live within the cells of ticks, mites, chiggers, and fleas. Bites from these pests can transmit *Rickettsiae* and cause diseases such as Rocky Mountain spotted fever and epidemic typhus.

HOW ARE MITOCHONDRIAL GENES INHERITED?

Egg cells are much larger than sperm cells and may have up to ten thousand times as many mitochondria. When a sperm and egg cell unite, both contribute approximately equal amounts of DNA to the cell nucleus. However, because practically all of the mitochondrial DNA comes from the egg, we inherit our mitochondrial genes from our mothers. Scientists can trace pure maternal lineages, without interference of paternal contributions, by analyzing mitochondrial DNA.

Because it is passed on in a maternal fashion, mitochondrial DNA does not mingle with DNA contributed by the father. The only source for variation in mitochondrial DNA is mutation. You might expect variation in mitochondrial DNA to be much less than variation in nuclear DNA, where recombination of maternal and paternal DNA, as well as mutation, contribute to variation. Surprisingly, that is not the case. On a percentage basis, mitochondrial DNA varies considerably more because it has a higher mutation rate. The nucleus has proteins that efficiently repair DNA when it is damaged, preventing most but not all mutations. The DNA-repair mechanisms of mitochondria are more primitive and are not as effective, so the mutation rate is higher.

For these reasons, mitochondria offer one of the best opportunities to compare human populations, compare species, reconstruct ancestries, and determine evolutionary relationships. As you will read later, mitochondrial DNA helped scientists identify the place where humans originated and reconstruct ancient histories of human migration.

MITOCHONDRIAL DNA SEQUENCES SHED LIGHT ON THE TRICHOTOMY PROBLEM.

By the mid-1990s, mitochondrial genomes from indigenous human populations in different parts of the world had been fully sequenced, as had the mitochondrial genomes of common chimpanzees, bonobos (a smaller, relatively rare chimpanzee species confined to a small region of Zaire in Africa), gorillas, and orangutans. Comparisons could now begin in earnest.

In 1995, Satoshi Horai and his colleagues in Japan published a detailed analysis of these comparisons.[1] Not surprisingly, the species with the most similar sequences were the common chimpanzee and bonobo. The human mitochondrial genome turned out to be closest to theirs. The human-gorilla comparison revealed the same degree of similarity as the chimpanzee-gorilla comparison, and both were less similar than the human-chimpanzee comparison. The orangutan sequence diverged the most from the other four species.

Below is a family tree diagram based on the mitochondrial relationships of humans and the great apes that gives a distinct answer to the trichotomy problem: humans and chimpanzees are more closely related to each other than either is to gorillas.

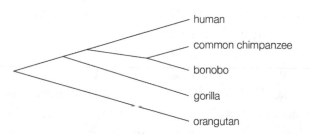

DO NUCLEAR DNA STUDIES
CONFIRM THE MITOCHONDRIAL ANSWER?

Long before complete mitochondrial DNA sequences were available, scientists were trying to solve the trichotomy problem by analyzing nuclear DNA. Most studies of individual genes in chromosomes suggested that chimpanzees and humans were the most closely related. The CMT1A duplication mentioned in chapter 2 is just one of many examples. But some studies suggested that chimpanzees and gorillas were more closely related, and a few even suggested the highest similarity between humans and gorillas. Because of the high degree of DNA similarity among all three species, as well as the natural variation inherent in nuclear DNA, statistical theory predicts that all three results would show up in isolated studies. Actual studies bear this out.

The differences are analogous to the signal-to-noise problem in broadcast television and radio. Before the days of cable and satellite television, people used either a small "rabbit-ears" antenna mounted on a television set or a large antenna on the roof. Random fluctuations in broadcast television signals are common, so reception rises and falls. A small antenna captures only a small part of the overall signal, so random fluctuations can readily cause the signal to fade in and out. The larger rooftop antenna captures a larger proportion of the signal. When the random fluctuations are averaged out, the signal is more consistent with less variation.

Because random variation and the laws of probability govern inheritance of DNA, we expect to see some random fluctuations, like the television noise, in DNA comparisons. A single isolated study is not always representative of an overall pattern. Although most studies may support a particular pattern, a few studies here and there might seem to contradict it because of the noise. Others might oversupport the real pattern and suggest a closer relationship than what actually exists. In other words, the noise varies in both directions.

However, when very large amounts of DNA sequence are compared, the random variations due to the noise are averaged out and the signal becomes much more apparent. For this reason, large studies on a genomic scale, which became available only recently, are much better indicators of evolutionary relationships than isolated small studies.

The concept of large-scale comparisons is not new. Charles Darwin, in his book *The Descent of Man*, wrote, "As we have no record of the lines of descent, the lines can be discovered only observing the degrees of resemblance between the beings which are to be classed. For this object, numerous points of resemblance are of much more importance than the amount of similarity or dissimilarity in a few points."[2]

The trichotomy problem persisted until genome projects generated enough information for large-scale comparisons of the DNA from humans, chimpanzees, and gorillas. Studies now include comparisons of genes, transposable elements, pseudogenes, and randomly selected sequences located between genes. All contribute to the big picture by averaging out the noise and making the signal clear.

In 2000, Satta et al. compared 45 chromosomal genes covering 46,855 base pairs and determined that the human-chimpanzee relationship is closest in the trichotomy.[3] A study by Wildman et al. published in 2003 covered 97 genes encompassing about 90,000 base pairs of sequence information. It strongly supports the human-chimpanzee relationship as the closest. The findings prompted the authors to propose reclassification of chimpanzees into the genus *Homo*, together with humans and extinct humanlike species.[4] Shi et al., also in 2003, examined 127 genes on chromosome 21 and determined that their research "unambiguously confirms the conclusion that chimpanzees were our closest relatives to the exclusion of other primates."[5] In 2003, Salem et al. looked at 133 *Alu* elements and determined that these elements point to the human-chimpanzee relationship as the closest.[6] In 2001, Chen and Li compared genes, *Alu* elements, pseudogenes, and randomly selected DNA sequences found in the regions of DNA located between genes. Their evidence clearly supports the human-chimpanzee relationship as the closest.[7] Finally, in 2004, Uddin et al. looked not at DNA sequences but at which genes function in the brains of humans, chimpanzees, and gorillas. They did so because human brain activity is cited as perhaps the most distinguishing human feature when compared to great apes. Not surprisingly, they found that the highest activity of genes in the brain was in humans. However, they also found more similarity between the genes active in the human and chimpanzee brains relative to the gorilla brain.[8]

Exceptions to the overall pattern have turned up in this research—the noise interfering with the signal. However, the majority of comparisons in each study support the human-chimpanzee relationship as the closest, in complete agreement with mitochondrial genome comparisons. The signal is now loud and clear, with a straightforward solution to the trichotomy problem: humans and chimpanzees are more closely related to each other than either is to gorillas.

NOTES

1. S. Horai et al., "Recent African Origin of Modern Humans Revealed by Complete Sequences of Hominoid Mitochondrial DNAs," *Proceedings of the National Academy of Sciences, USA* 92 (1995): 532–36.

2. C. Darwin, *The Descent of Man, and Selection in Relation to Sex*, 2nd ed. (London: John Murray, 1882), p. 148.

3. Y. Satta, J. Klein, and N. Takahata, "DNA Archives and Our Nearest Relative: The Trichotomy Problem Revisited," *Molecular Phylogenetics and Evolution* 14 (2000): 259–75.

4. D. E. Wildman et al., "Implications of Natural Selection in Shaping 99.4% Nonsynonymous DNA Identity between Humans and Chimpanzees: Enlarging the Genus *Homo*," *Proceedings of the National Academy of Sciences, USA* 100 (2003): 7181–88.

5. J. Shi et al., "Divergence of the Genes on Human Chromosome 21 between Human and Other Hominoids and Variation of Substitution Rates among Transcription Units," *Proceedings of the National Academy of Sciences, USA* 100 (2003): 8331–36.

6. A.-H. Salem et al., "*Alu* Elements and Hominid Phylogenetics," *Proceedings of the National Academy of Sciences, USA* 100 (2003): 12787–91.

7. F.-C. Chen and W.-H. Li, "Genomic Divergence between

Humans and Other Hominoids and the Effective Population Size of the Common Ancestor of Humans and Chimpanzees," *American Journal of Human Genetics* 68 (2001): 444–56.

8. M. Uddin et al., "Sister Grouping of Chimpanzees and Humans as Revealed by Genome-wide Phylogenetic Analysis of Brain Gene Expression Profiles," *Proceedings of the National Academy of Sciences, USA* 101 (2004): 2957–62.

Chapter 5

DARWINIAN DNA

The first name that comes to mind in connection with evolution is Charles Darwin. In fact, his name is so closely tied with evolution that some people incorrectly equate the words *Darwinism* and *evolution*. More than anyone else, Darwin focused the world's attention on the idea that changes occur in lines of descent—that life evolves—but he was not the first to propose it. We even find hints of evolutionary thought in the writings of Greek philosophers. However, for most of the past two thousand years, a literal interpretation of Genesis led people in Western societies to firmly believe that God specially and separately created species and that each species has changed little since the time of creation. Such literal interpretation is known as the *doctrine of special creation*. John Ray, a widely read seventeenth-century English author, expressed this doctrine when he referred to plants and animals as "the Works created by God at the first, and conserv'd to this day in the Same State and Condition in which they were first made."[1]

By the late eighteenth and early nineteenth centuries, several naturalists, observing the obvious hierarchical organization of life, broke ranks with the adherents of special creation and proposed that different lineages had diverged from common ancestors in the distant past and had eventually changed into new species. Perhaps the most prominent of these early evolutionists was Jean-Baptiste Lamarck, a French naturalist. He claimed that as individuals struggle to survive, their body parts change to help them adapt (like human muscles growing larger in response to weight lifting), and those changes are then inherited by their offspring. His idea, that environmental changes mold hereditary changes, is known as the *inheritance of acquired characteristics*, and Lamarck thought that it was powerful enough to produce new species. Although we now reject this notion, Lamarck's writings on the ability of species to evolve were very influential because they contradicted the doctrine of special creation, proposing common ancestry of separate species.

By the mid-nineteenth century, Darwin had realized that life evolves by way of *natural selection*. Individuals in a species vary in their characteristics, and many of those variations are inherited. An individual whose characteristics are favorable for survival and reproduction is more likely to pass on those characteristics to offspring than one whose characteristics are less favorable. Darwin's observations led him to believe that entirely new species could arise through natural selection.

It took Darwin years to figure out how life evolves. During a global voyage aboard a ship named the *Beagle*, he compiled notes identifying diverse organisms and their habitats in South America and on remote islands. He read the available literature and conducted hundreds of experiments. He was in the process of refining his theory of natural selection when Alfred Wallace independently came up with the same idea in 1858 and published it in a brief letter. Both Darwin and Wallace ended up sharing the credit but it was Darwin's book of about four hundred pages that rocked the scientific community. He titled it, with typical nineteenth-century verbosity, *On the Origin of*

Species by Means of Natural Selection or the Preservation of Favoured Races in the Struggle for Life, which we now simply call the *Origin of Species*. The book struck the world like a storm; the first printing sold out the same day it was released. It since has become one of the world's most influential books and it set the stage for a dramatic shift in scientific thought.

Darwin's and Wallace's theory of evolution by natural selection has undergone significant refinement over the years. Even so, it remains the unifying foundation of modern biology, well worth a quick glimpse of the reasoning that went into it.

Darwin set forth his main ideas in the first four chapters of the *Origin of Species*. In chapter 1, "Variation under Domestication," he describes how breeders have successfully produced dramatic changes in domesticated animals and plants. Breeders practice *artificial selection*: they select those individuals with desired characteristics and allow them to mate with one another, then repeat the process generation after generation. Darwin gives many examples of successful artificial selection, then argues that if humans can change organisms by selection, nature must be capable of doing so as well. There is a caveat: for selection to work, there must be variation, and the traits that vary must be heritable.

Chapter 2, "Variation under Nature," documents the variation within species that Darwin had personally observed or researched in the literature. He argues that variation and inheritance are absolutely essential for selection to be effective. He contends that if there is no inherited variation, neither nature nor breeders can preferentially favor one variant over another. Furthermore, even if there is variation for physical traits, those traits cannot be favored and successfully preserved unless they are passed from parents to offspring through inheritance.

In chapter 3, "Struggle for Existence," Darwin reinforces the point that individuals of all species are capable of producing more offspring than the environment can support. As a consequence, they must compete for food, shelter, mates, and other resources, and not all of them will survive to reproduce. Darwin carefully documented his field observations of individual plants and animals locked in a struggle for existence, in competition with one another for available resources. Think about the number of acorns an oak tree produces in its lifetime and the exceedingly few, if any, that successfully germinate and grow to become mature, acorn-producing trees. Individual plants and animals also differ in how successfully they elude predators, parasites, and disease. In nature, most die before reproducing.

Darwin then makes a concise, almost poetic inference that is central to his hypothesis:

> Owing to this struggle for life, any variation, however slight and from whatever cause proceeding, if it be in any degree profitable to an individual of any species . . . will tend to the preservation of that individual, and will generally be inherited by its offspring. The offspring, also, will thus have a better chance of surviving, for, of the many individuals of any species which are periodically born, but a small number can survive. I have called this principle, by which each slight variation, if useful, is preserved, by the term of Natural Selection, in order to mark its relation to man's power of selection. We have seen that man by selection can certainly produce great results, and can adapt organic beings to his own uses, through the accumulation of slight but useful variations, given to him by the hand of Nature. But Natural Selection, as we shall hereafter see, is a power incessantly ready for action, and is as immeasurably superior to man's feeble efforts, as the works of Nature are to those of Art.[2]

Darwin has now set the stage for chapter 4, "Natural Selection," the key chapter of his book. He cites many examples of variations in nature that natural selection can explain. For instance, natural selection can explain the differences among varieties in nature. Darwin understood a *variety* (which is now often called a subspecies) as a group of individuals with unique characteristics that distinguish them from individuals in other groups within the same species. Varieties in nature usually occupy specific geographic regions and members of the same variety preferentially mate with one another. Geographic barriers (such as rivers, canyons, mountain ranges, or oceans) typically isolate them from other varieties of the same species. Furthermore, the physical characteristics of a variety uniquely suit its members to the region where they live. For example, Mexican and Canadian varieties of coyotes are the same species, but the Canadian varieties tend to be much larger than the Mexican varieties. Large bodies allow better heat retention and are better in colder climates, whereas smaller bodies allow better heat dispersion and are favorable in warmer climates. Thus, the difference in body size between Canadian and Mexican varieties of coyotes makes sense when their native environments are considered in light of natural selection.

The key distinction between varieties and species of plants and animals is the ability to mate and produce fertile offspring. According to the classic definitions accepted in Darwin's day, members of different varieties of the same species can mate with each other and produce fertile offspring. Members of different species, by contrast, typically cannot produce offspring if they mate, or in rare cases they produce infertile offspring, such as mule (the offspring of a cross between a horse and a donkey), which is infertile.

Darwin proposed that natural selection's ability to produce varieties in nature ultimately results in the continued divergence of varieties until they become separate species. The remaining chapters in his book expound on specific evidences for the role of natural selection in the origin of species, including the roles of variation, geographic distribution, hybridization, and the necessity of long periods of time for new species to fully diverge.

Darwin knew little about the mechanisms of heredity, only that inheritance of variations was essential for his theory to work. Unfortunately, he was completely unaware that Gregor Mendel, an Augustinian monk, had recently discovered the basic principles of inheritance. The synthesis of Darwin's theory of natural selection and Mendel's theory of heredity did not happen until the 1930s. The combined Darwinian/Mendelian theory of natural selection is now known as neo-Darwinism and is the basis of our current understanding of natural selection.

Neo-Darwinism of the 1930s was destined to undergo large-scale expansion when the molecular basis of heredity was revealed. From the 1940s to the present, the sometimes gradual, sometimes lightning-paced discoveries in molecular biology have been filling in the details of neo-Darwinism, offering powerful evidence that Darwin's and Wallace's idea of natural selection is indeed a major key to evolution.

NATURAL SELECTION AFFECTS DNA SEQUENCES.

A basic tenet of biology is that genes encode proteins and proteins govern the development of inherited traits. Variation in the DNA sequences of genes often expresses itself as variation in proteins, which, in turn, causes variation in the traits governed by those genes and proteins. Ultimately, variation in DNA is the raw material for natural selection.

Mutations, recall, are the original source of variation in heritable traits. With respect to selection, mutations are either *selectively relevant* or *selectively neutral*. Selectively relevant mutations alter protein function, often by changing the protein itself. In most cases, a selectively relevant mutation reduces or eliminates the protein's function. If the protein contributes to survival and reproductive success, altered forms tend to be selected against. In time they disappear from the species.

In rare but important instances, a mutation alters the protein to make it *more* beneficial to the organism than it was before. In such a case, the mutation is favored and may eventually, over many generations, become established in the DNA of every individual in the species. Such a mutation is also selectively relevant.

Alternatively, a mutation may have no effect whatsoever on the function of a protein, so there is no selective advantage associated with its preservation or removal. The mutation is selectively neutral.

Some selectively neutral mutations are found in DNA segments that hold mainly useless information. Recall that a gene is a segment of DNA that encodes a protein. However, it does not encode the protein directly. Instead, the cell copies the DNA sequence into another type of molecule called RNA, which is very similar to DNA. The copied sequence of RNA then specifies the protein:

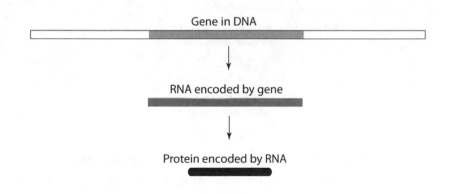

This concept of information transfer,

$$DNA \rightarrow RNA \rightarrow protein$$

is often called the *central dogma* of molecular biology and it explains much about evolution.

In the 1970s, when scientists compared the sequences of DNA in genes with the sequences of RNA encoded by those genes, they made a puzzling discovery: the DNA of most genes in animals, plants, and other eukaryotes contains *too much* information. The extra segments of largely useless information were named *introns*, and they must be cut out of RNA before the protein is made. *Exons* are the portions of the gene that remain in the RNA after the introns have been removed. Thus, we can expand the diagram of the central dogma to include intron removal and splicing of exons:

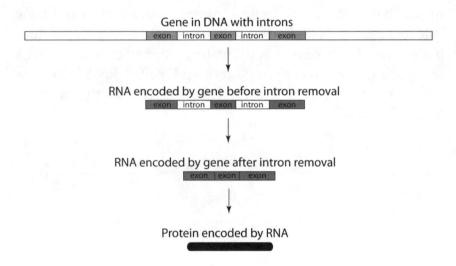

Several years ago I had an experience analogous to how introns are cut out of RNA. I was writing at my computer but left for a moment to run an errand. My three-year-old son wandered into the room and climbed onto the chair in front of the computer. When I returned, he was gleefully pounding on the computer keyboard, giggling as he watched hundreds of letters and numbers appear on the computer screen. I panicked, fearing that he had ruined hours of writing. I quickly escorted him from the room and managed to divert his interest to a pile of toys. I then frantically returned to the computer, fearing the worst. To my relief, he had not deleted anything, only inserted a lot of garbled letters and numbers. I selected the gibberish, hit the delete key, and the words I had written were once again connected as intended.

Suppose a gibberish-loaded sentence reads, "Charles Darwin published besiebrl6egbubgifnreu79dfbj his book, *The Origin of* yerygevfwuey4338 *Species*, in the ';lpko;j3678ghb0o year 1859." By cutting out the meaningless typing, we can restore the sentence to "Charles Darwin published his book, *The Origin of Species*, in the year 1859."

Introns are segments of DNA that contain mostly meaning-less sequences of bases like the gibberish in the sentence we just examined. They are not cut out of the DNA and thus are passed on from generation to generation, but they are cut out of the RNA after it has been copied from the DNA in a gene. As a result, most mutations in introns have no influence on the pro-tein specified by the gene, because the introns containing the mutations are removed before the protein is made. Therefore, most mutations in introns are selectively neutral, whereas most mutations in exons are selectively relevant.

Suppose we compare the DNA sequences of the same gene in two species that have common ancestry. Because selectively relevant mutations tend to be disfavored, we shouldn't find many in the exons. Conversely, we expect to see a larger pro-portion of mutations in introns because most of the mutations are selectively neutral. Three examples will reinforce the point.

TRANSPOSABLE ELEMENTS ARE OFTEN FOUND IN INTRONS, WHERE THEY ARE SELECTIVELY NEUTRAL.

In the summer of 2005, one of my colleagues and I sequenced the DNA of the chimpanzee *NANOG* gene, which becomes active soon after conception.[3] This gene has three introns, and two of them gave us some trouble when we tried to sequence them. Segments of DNA that contain repeats are more difficult to sequence than those with no repeats, and, as it turns out, the first intron in the *NANOG* gene has three *Alu* elements in it plus a fragment of a fourth.[4] The second intron has two *Alu* elements in it, plus fragments of two more. Not surprisingly, when we compared the chimpanzee and human *NANOG* genes, we found that both have all five *Alu* elements, and all three fragments of *Alu* elements, in the same places and the same orientations. However, in both chimpanzees and humans, the gene's protein-encoding portion has no *Alu* elements. This pattern is common. Transposable elements are often found in introns but not in the protein-coding parts of exons.

There is no evidence that introns are like magnets, drawing transposable elements into them. Transposable elements are just as likely to insert themselves into the protein-encoding parts of a gene as any other part. But when they do enter the protein-encoding region, they disrupt the gene's function, often with dire consequences such as cancer or a genetic disorder.

A transposable element stuck in the protein-encoding part of a gene is a selectively relevant mutation, usually one that adversely affects survival and reproduction. Eventually most of these disrupted genes are removed by way of natural selection, so we don't often find transposable elements in the essential protein-encoding parts of genes. When we do come across them, we find that they have recently inserted themselves there and the mutated gene is still being selected against in the population at large.

By contrast, a transposable element stuck in an intron is selectively neutral because the cell removes the intron, along with the transposable element, from the RNA before it makes the protein. The transposable element can persist generation after generation, even for millions of years, while the gene continues to function normally.

THE *INS* GENE IS AN EXCELLENT EXAMPLE
OF THE EFFECT OF SELECTION ON DNA SEQUENCE.

Let's turn now to the *INS* gene, which encodes insulin, a protein that stimulates cells to extract glucose as a source of energy from the blood. Most people have heard about insulin as a treatment for diabetes. Diabetics either don't produce enough insulin or their cells don't respond to insulin as well as they should. In either case, insulin injections often help them overcome many of the ill effects of diabetes.

Most mutations in the protein-encoding part of the *INS* gene are selectively relevant; they cause severe diabetes in early childhood. Very few people carry a selectively relevant mutation in this gene. Although diabetes is common, most cases are *not* due to a mutation in the *INS* gene and most of them appear during adulthood, after people have reproduced. Throughout most of human history, insulin treatments for diabetes were unknown and, therefore, unavailable. Children who were born with severe diabetes because of a mutation in the *INS* gene usually died before reproducing. The mutations were lost from the human population.

Now let's compare the *INS* gene in humans and chimpanzees. The insulin proteins of humans and chimpanzees are 100 percent identical. Their *INS* genes, however, are not quite identical; 97.7 percent of the bases are the same. They differ in twenty-nine bases, twenty-three within the introns. The other six mutations are in exons but have no effect on the protein. Thus, all of the twenty-nine mutations that distinguish the human and chimpanzee *INS* genes are selectively neutral.

As we've seen, humans and chimpanzees are closely related species. Suppose we compared the human *INS* gene with its counterpart in a more distantly related mammal—say, in pigs. Would we still see a pattern explainable by natural selection? We expect to see a larger number of mutations distinguishing the *INS* genes of these two species, yet the insulin proteins of humans and pigs are almost identical. The diagram below compares their *INS* genes at the boundary between essential protein-encoding sequence in an exon and the noncoding sequence in an intron. Wherever there is a vertical line, the two sequences match.

Human sequence
GGCTTCTTCTACACACCCAAGACCCGCCGGGAGGCAGAGGACCTGCAGGGTGAGCCAACTGCCCATTGCTGCCCCTGGCCGCCCCCAGCCACCCCCTGCTCC
|||||||||||||| |||||| |||| |||||||| ||| ||| |||||||||| | | || | | | | | |
GGCTTCTTCTACACGCCCAAGGCCCGTCGGGAGGCGGAGAACCCTCAGGGTGAGCCGAGGGGGCGTCCCGGGAGCGGTCGGGGGAGTTTTTAAGGAGGAAAT
Pig sequence   ◄——— exon |intron ———►

Within the exon, the pig and human genes are highly similar, but the similarity drops off precipitously within the intron. We expect the essential exon sequences to be preserved as selectively relevant mutations are eliminated, while selectively neutral mutations accumulate in the introns over evolutionary time. This is exactly the pattern we see in this gene and, in fact, just about every other gene. Interestingly, the first six bases on the end of this intron target the intron for removal, so mutations in this small part of the intron *are* selectively relevant. These six bases are preserved in both the human and the pig versions.

PSEUDOGENES ALLOW US TO VERIFY
THE EFFECT OF NATURAL SELECTION ON GENES.

When genes from different species are compared, they consistently display the same pattern: DNA sequences within exons are highly similar, and those within introns are less so. How do we know the pattern is evidence of natural selection? Here's one test. If two species have the same pseudogene in the same place with the same inactivating mutations, then the pseudogene must have been inactivated before the divergence from their common ancestral lineage. Pseudogenes, recall, are completely inactive. Thus, any mutations that distinguish the same pseudogene in the two species should be selectively neutral regardless of whether they are in the pseudogene's exons or introns. If mutations are disfavored in the exons of active genes but not in pseudogenes, there should be no difference in the proportion of mutations in the exons and introns of a pseudogene.

Reflect on the *GULO* pseudogene we discussed in chapter 3. It mutated into a pseudogene in a common ancestor of all primates, so all modern primates have it. If the agents of selection act on genes but not pseudogenes, we should see no difference in the proportion of mutations in exons compared to introns between any two primate species.

Let's first compare the human and chimpanzee versions. Their introns and exons are 98 percent identical, as they are in nearly all human and chimpanzee pseudogenes. Exons in most active genes, however, are more than 99 percent identical in humans and chimpanzees. Natural selection theory can explain the slight increase in similarity as preservation of essential sequences.

The difference should be even greater if we examine more distantly related species. The rhesus macaque is an Old World monkey species, less closely related to humans than chimpanzees. When we compare the human and the rhesus macaque *GULO* pseudogenes, the similarity is the same in exons and introns: 92 percent in the exons and 92 percent in the introns— as expected for selectively neutral mutations.[5]

So far, we've looked at powerful evidence of human evolution in chromosomes, transposable elements, pseudogenes, and genes. In each case, we've examined a few straightforward examples. Completion of the human genome and the first draft of the chimpanzee genome opened the door to literally millions of similar examples, overwhelmingly confirming our evolutionary history. In the next chapter, we turn to these two genomes to examine the evidence of evolution on a genomic scale.

NOTES

1. J. Ray, *The Wisdom of God Manifested in the Works of Creation* (London: R. Harbin at the Prince's-Arms in St. Paul's Church Yard, 1717), preface.

2. Ibid.

3. D. J. Fairbanks and P. J. Maughan, "Evolution of the *NANOG* Pseudogene Family in the Human and Chimpanzee Genomes," *BMC Evolutionary Biology* 6 (2006): 12, http://www.biomedcentral.com/1471-2148/6/12.

4. We deposited the sequence of this gene in the GenBank DNA database, the major worldwide database for DNA sequence information, and it is freely available over the Internet. To access it, go to http://www.ncbi.nlm.nih/gov. In the first search field, choose "Nucleotide" for the database. In the second search field (the one after the word "for"), type in "DQ179631," which is the accession number for the sequences, and click "Go." Then click on the link to the number and the information about this gene should appear. Scroll down past the information to the bottom and you will see the DNA sequence of this gene.

5. The comparison is based on a BLAST search, http://www.ncbi.nlm.nih.gov, I conducted with the *Homo sapiens* and *Macaca mulatta* versions of the *GULO* pseudogene on July 1, 2006.

Chapter 6

A SPECTACULAR CONFIRMATION

Of all animals on Earth, chimpanzees are most like us. For years, people thought that humans were the only species to use tools and practice organized tribal warfare. However, Jane Goodall, who has devoted most of her life to studying chimpanzees in the wild, discovered that they practice these and many other humanlike behaviors.[1] Even so, our spectacular analytical abilities, verbal skills, and cultural and technological innovations put us in a class by ourselves.

Hundreds of studies have revealed how remarkably close humans and chimpanzees are in their genetic makeup. Thousands of genes are almost identical, the chromosomes are highly similar, and transposable elements and pseudogenes are usually found in the same places. But to truly understand the degree of genetic relationship, we need to compare the entire genomes of both species.

September 2005 was a banner month for science when the Chimpanzee Sequencing and Analysis Consortium, a group of more than sixty scientists spread across several institutions, published the first draft of the DNA sequence of the chimpanzee genome.[2] They introduced it with these lines: "More than a century ago, Darwin and Huxley posited that humans share recent common ancestors with the African great apes. Modern molecular studies have spectacularly confirmed this prediction." Although they were referring to hundreds of earlier studies, the same can be said of their work: it is a spectacular confirmation of our recent shared ancestry with chimpanzees.

Beyond such confirmation (which was not even in doubt by then), sequencing of the human and chimpanzee genomes opened the door to an almost incomprehensibly vast opportunity to compare not just selected examples but our entire genome with that of our closest living evolutionary relative. In this chapter, we'll do just that, returning to the topics of previous chapters—chromosomes, transposable elements, pseudogenes, and genes—but this time we'll look at them on a genomic scale to see how close the human and chimpanzee genomes are and, more important, how they differ.

HUMAN AND CHIMPANZEE CHROMOSOMES DIFFER BY TEN MAJOR REARRANGEMENTS.

Long before genome sequencing zoomed into the research stratosphere, microscopists had been comparing human and chimpanzee chromosomes. They discovered that human chromosomes perfectly aligned with chimpanzee chromosomes except in a few places, some quite obvious. Later, genome comparisons confirmed the alignments in far greater detail. In figure 6.1, all of the aligned regions are coded gray and all regions that differ are coded white. Major differences show up in ten regions. One of them is the fusion site and lost centromere in human chromosome 2, as explained in chapter 1. The other nine are *inversions*, where a segment's orientation is reversed in one species when compared to its corresponding segment from

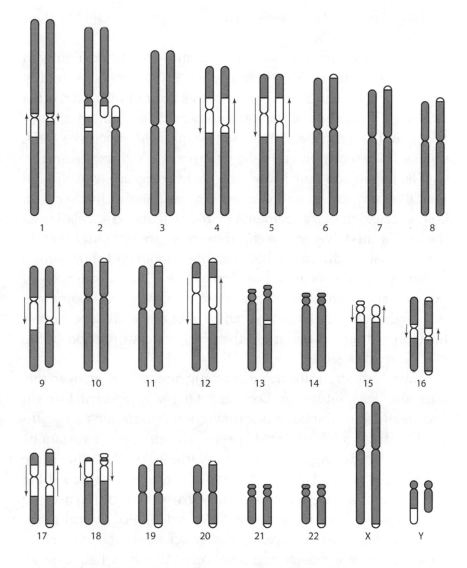

Figure 6.1. Alignment of human and chimpanzee chromosomes. The human chromosomes are on the left and the corresponding chimpanzee chromosomes are on the right. Regions in gray directly align with each other; those in white either do not align or align when rearranged. The arrows depict inversions. An arrow pointing downward represents the original orientation and the arrow pointing upward represents the segment that was inverted.

the other species. The reverse arrows in figure 6.1 identify their locations.

Human and chimpanzee chromosomes also differ slightly in a few other places. Most are expanded segments of highly repeated DNA sequences in one species but not the other, often on one or both ends of the chromosomes. The human Y chromosome, for example, has a much-expanded repetitive DNA segment when compared to the chimpanzee Y chromosome.

The fusion and nine inversions are *rearrangements*. Although the DNA sequences within them are essentially the same in both species, the arrangements of those sequences, relative to the rest of the DNA in the chromosomes, are not. And therein lies the greatest difference between the human and chimpanzee genomes. It is sometimes said that the DNA sequences of humans and chimpanzees are more than 98 percent similar.[3] Although true, this statement can be a bit misleading because it fails to point out that a more-than-insignificant portion of the DNA is rearranged.

How important are these rearrangements? To answer this question, let's return to Darwin's *Origin of Species*. Darwin spends much of the book discussing what variations are sufficient to distinguish different species, as opposed to variations that simply distinguish different varieties within the same species. All he had available at the time for distinguishing species and varieties were the outward, visible differences he could observe and, in some cases, knowledge of hybrid infertility. A little more than a century later, scientists observed chromosome rearrangements and discovered that they help explain the issue that so captivated Darwin: how varieties of the same species could eventually become different species.

For two new species to diverge from a single ancestral species, their ancestors must first be reproductively isolated from each other. This means that something separates the two ancestral groups—usually a geographic barrier, such as an ocean or mountain range—so that there is no intermating between them. Once such isolation has been established, any genetic changes that happen in one group are completely independent of those that happen in the other group. Over time, generation after generation, genetic changes accumulate independently in the two groups, causing them to gradually diverge from one another.

Ultimately, if they remain isolated from each other long enough, the two groups diverge so much that if brought back together they can no longer mate with each other and produce fertile offspring. At that point, they are considered as fully separate species. What causes the loss of interfertility when species diverge? There are several genetic causes, but chromosomal rearrangements are some of the most important. Most mutations do not cause any loss of fertility, but chromosome rearrangements do. A single inversion on a chromosome inherited from one of the two parents causes a slight loss of fertility. So does a fusion. In fact, single chromosome rearrangements like inversions and fusions are a major cause of infertility in humans. Although one chromosome rearrangement is usually insufficient for complete loss of fertility, multiple rearrangements can result in complete sterility, as in mules.[4]

Therefore, if we compare closely related species that at one time had common ancestry but have since diverged independently over time, we ought to find multiple chromosome rearrangements when we compare their chromosomes. Such is the case when we compare human and chimpanzee chromosomes, which differ by ten major rearrangements. In fact, the human and chimpanzee genome projects revealed even more inversions that distinguish the two species, inversions that are too small for scientists to observe with a microscope. The same pattern is evident between many other closely related species. For example, the chromosomes of mice and rats, or horses and donkeys, show considerable similarities but differ with respect to rearrangements. Chromosome rearrangements have now been examined in enough species of apes, monkeys, and other primates for scientists to reconstruct the original chromosome arrangement in their common ancestor.[5]

RETROELEMENTS HAVE BEEN ACTIVE IN BOTH THE HUMAN AND THE CHIMPANZEE GENOMES SINCE THEY DIVERGED.

Transposons, the DNA elements that transpose through a cut-and-paste mechanism, mutated so much that they were no longer transposing when the human and chimpanzee lineages diverged. Therefore, we expect to find the overwhelming majority of transposons in precisely the same places in the two genomes—and we do. Some retroelements, however, are still active in humans and chimpanzees, so we expect to find some recent retroelement insertions that are unique to one genome or the other—and we do. Even so, more than 99 percent of the millions of retroelements in both genomes are in the same positions, confirming that the majority of all transposable element insertions preceded the divergence that gave rise to the human and chimpanzee lineages.

The small fractions of unique transposable elements in both genomes are clues to evolutionarily recent transposable element activity. Among the most fascinating are retroelements that resemble retroviruses. There is no evidence that retroviruses invaded the human genome since the divergence. However, two viruslike retroelements are unique to the chimpanzee genome where they have inserted a few hundred copies of themselves. Two retroviruses apparently infected the chimpanzee ancestral genome some time after the divergence and mutated into retroelements. Both are still actively retrotransposing.

The genome in our distant ancestors experienced a burst of *Alu* element insertions some time after the divergence. *Alu* elements were also transposing in the ancestral lineage of chimpanzees during the same time span but at a much slower rate. For example, in 2004, scientists found more than twenty-two hundred copies of the relatively recent *AluYb8* element in the human genome but only nine in the chimpanzee genome.[6] Also, the human genome has 1,452 copies of the recent *AluYa* element but the genomes of other primates have only five.[7]

Of the millions of *Alu* elements in the human and chimpanzee genomes, 7,082 are unique to humans, whereas only 2,340 are unique to chimpanzees.[8] In other words, *Alu* elements were about three times more active in the human ancestral lineage. Fortunately for us, the burst of activity is over and transposition is now much less active than it once was in our ancestry.

NEW PSEUDOGENES HAVE ENTERED THE ANCESTRAL GENOMES OF HUMANS AND CHIMPANZEES SINCE THEY DIVERGED.

There are nearly 20,000 pseudogenes in the human and the chimpanzee genomes. Like transposable elements, nearly all are located at exactly the same places. However, a few new pseudogenes have inserted themselves into both genomes since they diverged, 163 in humans and 246 in the chimpanzee genome, according to a recent count.[9]

Scientists have a simple but reliable way to estimate when a particular pseudogene inserted itself into the genome. The logic goes like this: When a gene generates a pseudogene copy of itself, the original gene and the copy should be identical. Because the pseudogene is not functional, any mutations it acquires after its insertion are selectively neutral. Because they are neither favored nor disfavored, the mutations should accumulate at a fairly constant rate. Therefore, the more a pseudogene differs from its original gene, the older it must be.

By this logic, if a pseudogene arose after the divergence of human and chimpanzee ancestral lineages, it should be present in either the human or chimpanzee genome but not both. Also, it should be more similar to the original gene than more ancient pseudogenes in both genomes.

Let's take a look at the *NANOG* gene and its pseudogenes as an example. Its name, a reference to the land of eternal youth in Celtic mythology (Tir-Na-Nog), is appropriate. The gene's product allows cultured embryonic stem cells to keep growing and dividing indefinitely, as if they had eternal youth. The medical implications are enormous; cultured embryonic stem cells have the potential to treat, and possibly cure, debilitating diseases such as diabetes and Alzheimer's disease.

The human and chimpanzee genomes each have ten *NANOG* pseudogenes, all in the same places, but the human genome has one more (figure 6.2). The extra one, *NANOGP8*, is located on human chromosome 15.[10] Scientists at Oxford Uni-

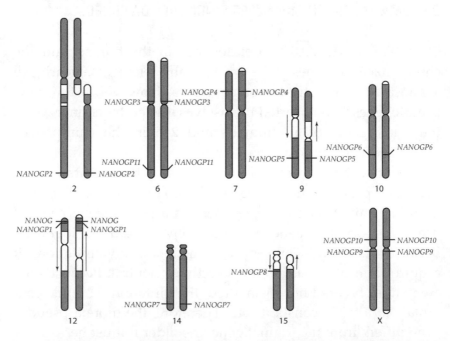

Figure 6.2. Positions of the *NANOG* gene and the eleven *NANOG* pseudogenes in human and chimpanzee chromosomes. Human chromosomes are on the left and chimpanzee chromosomes on the right. *NANOGP8* is absent from chimpanzee chromosome 15.

versity determined the relative ages of the eleven human *NANOG* pseudogenes by comparing the number of mutations in each with the functional gene.[11] By their analysis, *NANOGP8* is the most similar to the *NANOG* gene and therefore must be the youngest. They estimated its origin to be about five million years ago.

Another discovery confirms that *NANOGP8* is the youngest of the eleven pseudogenes. In both humans and chimpanzees, an *Alu* element is positioned near one end of the functional *NANOG* gene. Of all the pseudogene copies, only *NANOGP8* has this same *Alu* element. This *Alu* element must have inserted itself into the *NANOG* gene after the first ten pseudogenes originated but before the human and chimpanzee ancestral lineages diverged. After the divergence, the *Alu* element was already in the *NANOG* gene and was copied into *NANOGP8*, an independent line of evidence telling us that *NANOGP8* is the youngest of the eleven pseudogenes.

Nearly all studies on the divergence of human and chimpanzee ancestries place it between five and seven million years ago. Thus, if *NANOGP8* originated about five million years ago, it probably arose *after* the split between the human and chimpanzee ancestral lineages and chimpanzee DNA shouldn't have it. One of my colleagues and I searched for a site in chimpanzee DNA that corresponds to the site of *NANOGP8* in human DNA. We found the site, and a pseudogene isn't there.[12]

NANOGP8 pseudogene

── 2,123 bases ──
[CAAAGC......AAAAAA]

Human DNA	GCCTTCAAGCATCTGTTTAACAAAGCATATCTTGCCACCG
Chimpanzee DNA	GCCTTCAATCATCTGTTTAACAAAGCATATCTTGCCACCG

The findings of all studies into evolution of the *NANOG* pseudogenes are consistent. Ten pseudogenes arose before the divergence of human and chimpanzee ancestral lineages, and an *Alu* element inserted itself into the *NANOG* gene after these ten pseudogenes were formed but before the divergence. The *NANOGP8* pseudogene then inserted itself into chromosome 15 in the human ancestral lineage after the divergence.

NATURAL SELECTION EXPLAINS SOME OF THE DIFFERENCES BETWEEN HUMAN AND CHIMPANZEE GENES.

Let's now turn our attention to the core of any genome—the genes. Although they make up less than 2 percent of our genome, our genes encode the proteins to make us human. They are also the part of the genome most subject to natural selection because mutations in genes create new variations that may prove either useful or harmful to the individual. Therefore, natural selection should explain some, if not most, of the major differences between human and chimpanzee genes.

When the genes in both genomes are compared, the pattern is the same as that for the rest of the genome—the same genes are in the same places in the chromosomes and their DNA sequences are highly similar, in fact even more similar (more than 99 percent in most cases)[13] than DNA sequences elsewhere. Remarkably, 29 percent of human and chimpanzee genes encode exactly the same proteins. Most of the others differ by very small amounts, usually in regions that have little or no effect on function.

Recent ancestry combined with natural selection explain this higher similarity within genes. In their structure and physiology, humans and chimpanzees have much in common. Therefore, the products of most genes are subject to the same or similar environmental constraints. Most selectively relevant mutations in either species should be selected against and the original sequence conserved. For example, insulin carries out the same vital role of blood sugar regulation in both humans and chimpanzees. As we saw in the previous chapter, human and chimpanzee versions of the insulin protein are identical. All mutations that distinguish the two genes are in areas that have no effect on the protein. The same can be said for thousands of other human and chimpanzee genes.

Let's focus instead on examples of genes that differ between humans and chimpanzees. Genes that govern resistance to certain infectious diseases, including malaria and tuberculosis, are especially intriguing. Let's look at malaria as an example. When considering natural selection and infectious diseases, we must keep in mind that both the parasite and its host are living organisms and natural selection can explain variations in their genes, albeit in different ways. In parts of the world where malaria is prevalent, people who carry a mutation that confers resistance to malaria are more likely to survive and pass the mutated gene to offspring than those who do not. Over generations, the proportion of people who have the mutated gene increases in regions where malaria is prevalent—they have a selective advantage. However, the parasite that causes malaria, the microorganism *Plasmodium falciparum*, is also subject to agents of natural selection. Any mutation that helps the parasite overcome the resistance is also favored. Parasites that carry the mutation are more likely to multiply as they infect more people. The host species and the parasitic species coevolve, with changes in one exerting selective pressure on the other.

This antagonistic host-pathogen coevolution results in rapid evolution of the DNA sequences of genes that influence the infection process—in both the pathogen and the host. *Plasmodium falciparum* causes malaria exclusively in humans. A closely related species, *Plasmodium reichenowi*, infects chimpanzees. Just as humans and chimpanzees share common ancestry, so do the two species of parasites. Most likely, their divergence paralleled the divergence of the host species.

If this is true, we ought to see a marked difference in human and chimpanzee genes that influence *Plasmodium* infection. Indeed, some of the greatest differences are in these genes.[14] Consider the *GPA* gene, which encodes a protein that the *Plasmodium* parasite first recognizes when it infects humans. The parasite has its own gene, *EBA-175*, which encodes a protein that attaches to the *GPA* gene's protein during infection. If *GPA* mutates to produce a protein that the *EBA-175* gene's protein can no longer recognize, anyone who has this mutant version of the *GPA* gene should be resistant to malaria. On the other hand, suppose a chance mutation in the *EBA-175* gene changes the protein in such a way that it can now attach to the altered *GPA* gene's protein; the resistance is lost and the parasite can multiply.

This tit-for-tat mutating combined with selection should leave evidence of rapid change in the DNA sequences of both human and parasite genes. In 2003, researchers in Taiwan and the United States compared the *GPA* gene in humans and chimpanzees and found that it varied far more than most other genes, which is strong evidence of natural selection and rapid change.[15] The clincher, however, was their comparison of *EBA-175* in *Plasmodium falciparum* (the human parasite) and *Plasmodium reichenowi* (the chimpanzee parasite). In both parasites this gene varied more than other genes, evidence that *GPA* and *EBA-175* coevolved in the host-parasite pairs.

Other genes that are highly diverged between humans and chimpanzees are those that govern reproduction, encode proteins in hair, and an especially large number that influence the sense of smell. The divergence of these genes is expected when we consider natural selection. Reproduction is quite different in humans and chimpanzees, as is the advantage for body hair, and chimpanzees have a much more acute sense of smell than we do, because their survival depends on detecting odors to find food and identify danger.

Of all the important differences between humans and chimpanzees, the one that most intrigues us is brain function. Our ability to speak with complex sentences, our use of written language, our technological development, our propensity to reason, and our overall advanced intellectual capacity are perhaps our most uniquely human traits, all due to our highly developed brains. Thus, we should expect to find some marked differences between humans and chimpanzees in the genes that govern brain development. Scientists at the University of California, Santa Cruz, compared the human and chimpanzee genomes with those of other vertebrates, searching for regions that displayed an unusually high number of mutations in the human genome. They called these regions human accelerated regions (HARs) and they found 202 of them.[16] Most HARs do not contain genes but instead contain DNA sequences that regulate how, when, and to what extent a gene is turned on or off. In other words, differences in gene regulation rather than in the genes themselves determine much of the difference between humans and chimpanzees. Genes known to govern brain development are especially prone to be close to a HAR.

One of their discoveries was especially important. The one HAR that the authors identified as showing "the most dramatically accelerated change" contains a previously unidentified gene that they named *HAR1F*.[17] This gene does not make a protein, only an RNA, and it is turned on in the embryo's developing brain from seven to nineteen weeks after conception, one of the most crucial periods for brain development. The human gene is highly different from the chimpanzee version. Of 118 base pairs in it (it is a very small gene), 18 differ between the human and chimpanzee versions, a difference of 15.3 percent, compared to the genome-wide difference of less than 2 percent. Comparison with the same region in gorillas, orangutans, and rhesus macaques reveals that all 18 of these mutations arose exclusively in the human lineage. To what extent, if any, does this small but uniquely human gene explain the substantially advanced brain development in humans? Did natural selection favor the 18 mutations in humans? So far, there are no definitive answers; we will need to wait for the results of further research.

In sum, genome-wide comparison of the human and chimpanzee genomes "spectacularly confirms" what previous individual studies have shown: the genes, chromosomes, transposable elements, and pseudogenes of humans and chimpanzees are strikingly similar. Although the molecular differences constitute only a fraction of the two genomes, they are not trivial. They represent some of the most powerful evidence of common ancestry because they are fully consistent with known mechanisms of chromosome rearrangement, generation of recent transposable elements and pseudogenes, and the effects of natural selection we expect to observe in certain genes and their regulatory regions. The comparison is massive, including thousands of genes and pseudogenes, millions of transposable elements, and billions of base pairs in DNA.

NOTES

1. For information about Jane Goodall, her work, and her discoveries, visit the Jane Goodall Institute Web site at http://www.janegoodall.org.

2. Chimpanzee Sequencing and Analysis Consortium, "Initial Sequence of the Chimpanzee Genome and Comparison with the Human Genome," *Nature* 437 (2005): 69–87.

3. According to the Chimpanzee Sequencing and Analysis Consortium (ibid.), the two genomes are 98.77 percent identical in DNA sequence.

4. F. Yang et al., "Refined Genome-wide Comparative Map of the Domestic Horse, Donkey and Human Based on Cross-Species Chromosome Painting: Insight into the Occasional Fertility of Mules," *Chromosome Research* 12 (2004): 65–76.

5. S. Muller et al., "Defining the Ancestral Karyotype of All

Primates by Multidirectional Chromosome Painting between Tree Shrews, Lemurs and Humans," *Chromosoma* 108 (1999): 393–400.

6. R. Gibbons et al., "Distinguishing Humans from Great Apes with *Alu*Yb8 Repeats," *Journal of Molecular Biology* 339 (2004): 721–29.

7. A. C. Otieno et al., "Analysis of the *Alu* Ya-lineage," *Journal of Molecular Biology* 342 (2004): 109–18.

8. Chimpanzee Sequencing and Analysis Consortium, "Initial Sequence of the Chimpanzee Genome."

9. Ibid.

10. D. J. Fairbanks and P. J. Maughan, "Evolution of the *NANOG* Pseudogene Family in the Human and Chimpanzee Genomes," *BMC Evolutionary Biology* 6 (2006): 12, http://www.biomedcentral.com/1471-2148/6/12).

11. H. A. F. Booth and P. W. H. Holland, "Eleven Daughters of *NANOG*," *Genomics* 84 (2004): 229–38.

12. Fairbanks and Maughan, "Evolution of the *NANOG* Pseudogene Family."

13. D. E. Wildman et al., "Implications of Natural Selection in Shaping 99.4% Nonsynonymous DNA Identity between Humans and Chimpanzees: Enlarging Genus *Homo*," *Proceedings of the National Academy of Sciences, USA* 100 (2003): 7181–88.

14. M. J. Martin et al., "Evolution of Human-Chimpanzee Differences in Malaria Susceptibility: Relationship to Human Genetic Loss of N-glycolylneuraminic Acid," *Proceedings of the National Academy of Sciences, USA* 102 (2005): 12819–24; Chimpanzee Sequencing and Analysis Consortium, "Initial Sequence of the Chimpanzee Genome."

15. H. Y. Wang et al., "Rapidly Evolving Genes in Human. I. The Glycophorins and Their Possible Role in Evading Malaria Parasites," *Molecular Biology and Evolution* 20 (2003): 1795–1804.

16. K. S. Pollard et al., "Forces Shaping the Fastest Evolving Regions in the Human Genome," *PLoS Genetics* 2, no. 10 (2006): e168, http://genetics.plosjournals.org.

17. K. S. Pollard et al., "An RNA Gene Expressed during Cortical Development Evolved Rapidly in Humans," *Nature* 443 (2006): 167–72.

Chapter 7

DIVERSITY

In the fictional world of novels and movies, scientists often are depicted as uncaring, even devious and fanatical. In the real world, most people look to scientists with hope that their discoveries will solve some of the world's most urgent problems. Countless scientists have taken that hope to heart and pursued their research with a passion to benefit humanity. Nicolai Vavilov, a Russian biologist whose career blossomed during the early part of the twentieth century, was one of them. Vavilov was well schooled in genetics and evolution, having studied these sciences in England. After returning to his native Russia, his training and research led him to a brilliant genetic discovery. His willingness to act on it ended up touching the lives of nearly every human being, and his legacy continues to do so. Powerful politicians opposed his work but Vavilov refused to renounce his commitment to it. He was arrested and spent the rest of his life in prison, where he died of maltreatment and starvation. By all accounts, Vavilov was a heroic global figure. So why is his story so rarely told?

The answer probably has to do with the nature of his work.

Among the most admired scientists are those who made extraordinary discoveries in medicine, people like Marie Curie, Louis Pasteur, and Jonas Salk. Vavilov didn't discover a miracle cure, a way to prevent disease, or a powerful new medical treatment. Instead, he pioneered a worldwide effort to protect the world's long-term food supply by recognizing the consequences of some basic principles of evolution.

Like most of his contemporaries, Vavilov knew that the world's major food crops had originated in specific places. Corn and beans, for example, originated in Mexico and Central America, potatoes in Bolivia and Peru, rice in Southeast Asia, and wheat, barley, and oats in the Middle East. From these regions, people carried their crops to various parts of the world, so that by the early twentieth century, the United States was the world's largest producer of wheat and corn, eastern Europe and Russia produced the majority of the world's potatoes, and rice had spread throughout the tropics.

Vavilov discovered a pattern in every food-plant species he examined, a pattern predicted by evolutionary theory and one that his observations repeatedly confirmed: *The greatest genetic diversity for a species is in the place where it originated*. He called the places where our food plants originated their *centers of origin*. His observation made sense. Throughout history as people carried seeds to other parts of the world, they took only a subset of the seeds available in the center of origin. As the seeds were dispersed, from one point to another, the amount of diversity dwindled with each step.

By the early twentieth century, scientists had combined Darwin's theories of selection with an understanding of genetics to breed new strains of crop plants, types of wheat, corn, rice, and other species that gave far greater yields than their unimproved predecessors. Most of the plant breeding took place in the more advanced nations of Europe and North America, far from the centers of origin. The new European and North American strains had little genetic diversity. Most of the diversity still remained in old, unimproved types grown by peasants in the centers of origin, places like Latin America, the Middle East, the Indian Subcontinent, Africa, and southeast Asia.

The genetic diversity of food crops found in the centers of origin is vital for the future of the world's food supply. For example, diseases afflict many of the world's crops with devastating consequences for humanity. The great Irish potato famine, caused by a fungus that infected potatoes throughout Ireland, is the most famous of many widespread outbreaks. Today, major crop epidemics are rare, largely because modern plant breeders find and introduce genes that confer resistance to plant diseases into the new strains of crops they develop. And where do they find these genes? To Vavilov, and to plant breeders throughout the world, the answer was clear: the genes are found in highly diverse plants growing in the centers of origin. These centers hold a gold mine of variation, one that is essential for securing the world's food supply.

Vavilov also recognized that the genetic diversity of food plants in these centers was in peril. The new strains that were revolutionizing agricultural production could also be its downfall. As they spread from Europe and North America to other parts of the world, they eventually came full circle, back to their centers of origin. There, peasant farmers, anxious to increase their production, were choosing to plant the new, genetically uniform strains instead of the old, less productive but genetically diverse ones. Vavilov knew that the genetic diversity would be lost forever unless someone intervened.

For two decades, he led more than a hundred expeditions to collect and preserve the diversity of the world's food supply. He and his colleagues gathered seeds from throughout the world of nearly every food crop imaginable: wheat, corn, rice, barley, oats, peanuts, tomatoes, squash, beans, peas, and many others. They deposited the seeds in a scientific institute in Leningrad, where they cataloged them and carefully preserved them, like precious treasures in a museum.

However, Vavilov's life and work were about to take a tragic turn. In 1927, Trofim Lysenko, an obscure Russian agricultural researcher, was the subject of a *Pravda* article titled "The Green Fields of Winter." The article's author, journalist V. Fedorovich, describing his first impression of the man, wrote, "Lysenko gives one the feeling of a toothache; God give him health, he has a dejected mien. Stingy of words and insignificant of face is he; all one remembers is his sullen look creeping along the earth as if, at very least, he were ready to do someone in." Fedorovich went on to contrast Lysenko's lowly physical appearance with his rise from a peasant upbringing to become an accomplished agricultural scientist. He concluded, "The barefoot Professor Lysenko now has followers, pupils, and experimental fields. He is visited in the winter by agronomic luminaries who stand before the green fields of the experiment station, gratefully shaking his hand."[1]

Lysenko swiftly gained favor with Joseph Stalin and began to preach his own unsupported notions, bashing the traditional genetics and evolutionary biology of the West. He taught that scientists could create new genetic variation in plants simply by manipulating the environments where plants were grown. Vavilov's efforts to collect genetic diversity were, in Lysenko's mind, a waste of time, even worse, a Western ploy designed to destroy Soviet agriculture. In a 1935 speech delivered to a congress of farmers and attended by Stalin and other government members, Lysenko branded Vavilov and other Soviet biologists as "kulak-wreckers and saboteurs" who "instead of helping collective farmers, they did their destructive business, both in the scientific world and out of it." At the end of the speech, Stalin stood and shouted, "Bravo, comrade Lysenko, bravo!"[2]

With Stalin's support, farmers, government officials, and a very small number of scientists soon praised Lysenko as a man of the people who would make Soviet agriculture the envy of the world. Vavilov led the opposition against Lysenko's pseudoscience. Lysenko retaliated, accusing Vavilov of espousing the bourgeois theories and pseudoscience of the West. In 1937, he launched a campaign that specifically named Vavilov as an enemy of the people. Vavilov continued to openly oppose Lysenko but could not deter official government sponsorship of Lysenkoism. In one of his last attempts to challenge Lysenko in 1939, Vavilov reaffirmed his resolve when he emphatically exclaimed, "We shall go to the pyre, we shall burn, but we shall not retreat from our convictions."[3] Shortly thereafter, in 1940, government police arrested Vavilov and several other Soviet scientists and imprisoned them for their refusal to follow Lysenko.

After Vavilov's arrest, his colleagues at the institute in Leningrad quietly continued his work. In the winter of 1941–42, during the horrific siege of Leningrad by Hitler's army, food ran out. Tens of thousands of people starved to death. Carefully cataloged packets of rice, wheat, corn, peanuts, peas, and other foods holding the genetic diversity collected by Vavilov surrounded the scientists in the institute. Recognizing the need to preserve that diversity for future generations, they made a pact among themselves that no one would eat the seeds. All of them suffered from severe hunger. Nine chose to die of starvation rather than consume the genetic diversity under their care. Vavilov also gave his life for his convictions. He died in prison of starvation and illness along with scores of his colleagues who likewise met their deaths in prison while remaining true to their resolve.

In 1964, with Khrushchev's forced resignation, Lysenko finally fell from power. Three years later, the institute where Vavilov's collections were housed was renamed the N. I. Vavilov Research Institute of Plant Industry in honor of its former director.[4] Vavilov's collections still benefit humanity as scientists today continue to utilize, augment, and research the genetic diversity that he collected and that others gave their lives to protect.

VAVILOV'S CONCEPT OF DIVERSITY AND ORIGINS ALSO APPLIES TO HUMANS.

Why highlight Vavilov and his colleagues in a book devoted to human evolution? One reason is to honor scientists whose convictions led them to sacrifice their lives for humanity. Another is to show the intellectual regression, even tragedy, that results when governments suppress well-supported science in favor of politically popular but scientifically flawed ideas. The main reason is that Vavilov's observations on diversity and origins have been repeatedly tested in humans and portray a clear pattern of origins and evolution.

Reflect on Vavilov's major concept: *the greatest genetic diversity of a species is found in the region where it originated*. Many species are endemic: their natural populations are restricted to an area that is easily identifiable as their center of origin. For example, the bristlecone pine (the longest-lived species on Earth) is restricted to six states in the western United States and that is where it originated. Koalas and kangaroos are endemic to Australia, their center of origin. However, some species have spread from their centers of origin to occupy much of the earth. Seagulls, rock doves, finches, and many other bird species are examples. So are diverse species of insects such as fruit flies and honeybees. Human populations also have dispersed to occupy much of the world. Vavilov's concept is most useful when applied to such wide-ranging species as a way of identifying their centers of origin.

Human diversity is measurable in a number of ways, including language, culture, and physical attributes. However, Vavilov's concept depends only on the degree of *genetic* diversity, and the best way to conclusively measure genetic diversity is to examine DNA diversity. What do studies of DNA diversity tell us about human genetic diversity and origins? Let's examine a few.

HUMAN MITOCHONDRIAL DNA DIVERSITY REVEALS
AN AFRICAN ORIGIN FOR HUMANS.

Just as mitochondrial DNA can be used to compare genetic diversity among species, it can also be used within a species to determine the degrees of relationship among individuals. Initially scientists focused on selected regions of human mito-chondrial DNA.[5] At the time, DNA-sequencing methods were too laborious and expensive for them to sequence the entire mitochondrial genomes of multiple people. The three initial studies found the highest degree of diversity in people indigenous to sub-Saharan Africa (the part of Africa south of the Sahara Desert).

However, these studies suffered criticism, largely because they examined only small regions of mitochondrial DNA. By 2000, DNA-sequencing methods were much more efficient and less expensive. That year scientists in Sweden and Germany published a study comparing the complete DNA sequences of the mitochondrial genome in people with worldwide geographic origins.[6] Like the previous studies, this one found the highest diversity in sub-Saharan Africa—*twice* as high as the rest of the world combined! According to analysis of mitochondrial DNA diversity in light of Vavilov's principle, modern humans—*Homo sapiens*—originated in sub-Saharan Africa.

HUMAN Y-CHROMOSOME DIVERSITY REVEALS
AN AFRICAN ORIGIN FOR HUMANS.

As mentioned earlier, mitochondrial DNA is inherited only through maternal lineages. Although women pass on their mitochondrial DNA to both their daughters and sons, only the daughters pass it on to future generations. By contrast, the Y chromosome displays the reverse pattern of inheritance; it is passed on exclusively through the paternal lineage from fathers to sons. Whereas maternal inheritance can be traced in mitochondrial DNA, paternal inheritance can be traced in the DNA of the Y chromosome.

According to several large-scale studies, the greatest diversity for Y-chromosome DNA is in sub-Saharan African populations. That finding likewise supports a sub-Saharan African origin for modern humans.[7]

However, the degree of diversity, and the differences among geographic regions for Y-chromosome diversity, are much less than those for mitochondrial DNA. A large study found that the average number of differences for Y-chromosome DNA in sub-Saharan Africans was almost one and a quarter times higher than in Europeans, and about one and a third higher than in Asians.[8] Although not as dramatic as the twofold higher diversity for mitochondrial DNA, the data still point to a sub-Saharan African origin for the Y-chromosome DNA.

What explains the discrepancy between the mitochondrial and Y-chromosome studies? First, mitochondrial DNA mutates more often than other types of DNA. Second, lower Y-chromosome diversity reflects Darwinian selection. Such selection is well documented in both the political and the genetic histories of human societies.

Genghis Khan provides us with a notorious example. He is well known for his tyrannical rampage throughout central Asia to establish the largest land empire in history. To understand how selection played a role in his empire, we need to remember that the success of selection is dependent not just on survival but also on reproduction. Genghis Khan's sexual exploits throughout the region he conquered are well documented. Furthermore, his male relatives, who inherited the same Y chromosome as he, ruled much of the region even after his empire fell, and they also had power for sexual exploitation. In 2003, a group of collaborating scientists in England, China, Italy, Pakistan, Uzbekistan, and Mongolia published research on Y-chromosome variation throughout Asia.[9] A particular Y chromosome is the most prevalent in the region of Genghis Khan's former empire, found in about 8 percent of all the men who live there. The pattern of distribution and the time of its origin extrapolated from the data both point to the time and the area of his reign, strong evidence that thirty million men today inherited his Y chromosome.

During their reproductive years, women can bear only so many children but men have the biological capacity to father many. This fact suggests that diversity for Y-chromosome DNA may be much less than diversity for mitochondrial DNA, especially in societies where politically powerful males were reproductively promiscuous. The genetic data combined with historical information bear this out.

HUMAN X-CHROMOSOME DIVERSITY REVEALS AN AFRICAN ORIGIN FOR HUMANS.

Because men (XY) inherit only a single X chromosome, they are more vulnerable to mutations in genes on the X chromosome. Unlike women (XX), they do not have a built-in redundancy to cover for a mutated gene. That is why men are more prone to the symptoms of X-related genetic disorders, such as red-green color blindness and hemophilia. That is also why traits associated with genes on the X chromosome are subject to somewhat different selection pressures.

So far, the most extensive study on DNA diversity in the human X chromosome was published in 2004 by a group of scientists at the University of Arizona.[10] They found the highest diversity in sub-Saharan African populations, about 1.8 times that of all other populations in the world combined. Here again we have evidence of a sub-Saharan African origin for humans.

DIVERSITY FOR DNA SEQUENCES IN ALL CHROMOSOMES REVEALS A SUB-SAHARAN AFRICAN ORIGIN FOR HUMANS.

Numerous large-scale studies on various types of DNA sequences in human chromosomes show the same unmistakable pattern as that shown by mitochondrial DNA and the X and Y chromosomes: the highest diversity is consistently present in people indigenous to sub-Saharan Africa. Examples include studies of multiple genes, *Alu* elements, other retroelements, simple mutations, and mutations in repeated DNA sequences. Regardless of what part of the human genome is examined or what type of DNA sequence is sampled, the findings point to modern human origins in sub-Saharan Africa.

STUDIES OF GENETIC DIVERSITY ALLOW SCIENTISTS TO RECONSTRUCT ANCIENT HUMAN DISPERSALS.

Long before DNA sequence analysis was available, the Italian geneticist Luigi Luca Cavalli-Sforza asked this question: "Can the history of humankind be reconstructed on the basis of today's genetic situation? This was the question I posed to myself over forty years ago. I made a personal bet it could be done, because I believed the theory of evolution gives us the key."[11]

Cavalli-Sforza has devoted his entire career to that bet. Given the tremendous amount of information he and many others have discovered, no one can reasonably dispute that he won the bet. He began by studying the distributions of blood types (such as types A, B, AB, and O) among indigenous populations in different parts of the world. In 1962, he and the British geneticist and statistician A. W. F. Edwards came up with statistical groupings of indigenous populations based on how closely they shared variations for particular blood groups. They compiled the data independently of geography to hierarchically group indigenous populations as most different to progressively most similar. Even though geographic origin was not factored into their analysis, the results placed people from the same continent in the same group. Furthermore, it revealed a general pattern of geographic dispersal. The most diverse groups were the Africans and Europeans. The least diverse were people native to North and South America, and they were most similar to Asians.

A basic principle of genetic diversity and dispersal has emerged and it coincides with Vavilov's concept of diversity. With each successive wave of dispersal, diversity should decrease because only a subset of individuals migrate from a population, taking with them progressively smaller portions of genetic diversity. People who leave one area for another should be genetically most similar to, but less diverse than, the population from which they most recently emigrated. Based on this principle, the data of Cavalli-Sforza and Edwards suggest that the ancestors of people native to the Americas emigrated from Asia.

This initial study was based on only a few genes. To be truly reliable, a study should be based on a large sample of genetic diversity. By 1984, Cavalli-Sforza and his colleagues had studied diversity for 110 genes and constructed a refined worldwide family tree showing the degrees of genetic relatedness of indigenous populations and their most plausible dispersal routes. As Cavalli-Sforza wrote, "The biggest difference in the tree is between Africans and non-Africans, once again reinforcing many paleoanthropologists' view that modern humans originated in Africa and later spread around the world."[12] The tree also portrays a dispersal out of Africa, splitting into groups that populated Europe, the Middle East, and Asia. Australian aborigines, Pacific Islanders, and Southeast Asians are most similar to one another and probably emigrated from a single ancestral population in South Asia. Indigenous North and South Americans are most similar to Northeast and Arctic

Asians, which points to dispersal from northeast Asia across the Bering Strait (anciently a land bridge) into the Americas.

As DNA-sequencing methods and genomic-scale analysis became available in the 1990s, scientists followed Cavalli-Sforza's pioneering studies with large-scale DNA studies of their own that dwarfed all previous studies. If the earlier studies had flaws because of their small scale or unintentional bias in gene selection, the new studies would have revealed them. As it turned out, they confirmed all major interpretations in the older small-scale studies, while refining a few details.

Let's examine two examples: mitochondrial DNA diversity and Y-chromosome DNA diversity. These are especially good choices because they represent pure maternal and paternal lineages, and they are the parts of the human genome that have been most extensively studied for diversity. The logic goes something like this. A mutation found in the mitochondrial DNA or Y-chromosomal DNA of a large number of people in various parts of the world probably happened in a common ancestor of those people a very long time ago. On the other hand, a mutation found in a small group of people in a limited geographic region probably happened in a common ancestor of those people, but much more recently. Old and new mutations remain linked to each other in the mitochondrial genome or the Y chromosome, so scientists can reconstruct the hierarchical pattern of mutations in people who are indigenous to particular regions. By doing so, they reconstruct ancient dispersal patterns.

Multiple studies of both mitochondrial DNA[13] and the Y chromosome[14] have been completed, and they show the same general pattern of ancient human dispersals, as depicted in figure 7.1 (p. 116). According to these patterns, before there were any modern humans (*Homo sapiens sapiens*) outside of Africa, people emigrated from Africa to the Middle East approximately sixty thousand years ago. Some of their descendants radiated into Europe, and North, Central, and South Asia.

From South Asia in what is now India, some moved into Southeast Asia and from there to Australia and the Pacific Islands. At least one group emigrated from Siberia and Northeast Asia across what is now the Bering Strait into North America. Their descendants spread southward to colonize North, Middle, and South America.

The map in figure 7.1 depicts only ancient dispersals, those that happened more than ten thousand years ago. More recent paths not indicated on the map include a Viking migration from Scandinavia to North America, dispersals to the Pacific Islands, and the large-scale migrations associated with colonialism, slavery, and immigration during recent centuries.

The National Geographic Society has initiated the Genographic Project, a long-term research effort that is substantially expanding the DNA database for indigenous people throughout the world to better define the histories and migration patterns of ancient humans. The project and its aims are summarized in a book titled *Deep Ancestry* by Spencer Wells,[15] the Genographic Project's director, and at the project's Web site, http://www.nationalgeographic.com/genographic.

Automated DNA sequencing methods coupled with the rapidly expanding database of human DNA sequences are helping scientists reconstruct ancient dispersal patterns with a high degree of confidence. These same methods also are confirming and refining our understanding of evolutionary relationships among other species in a great branching tree of life. And that is the topic of the next chapter.

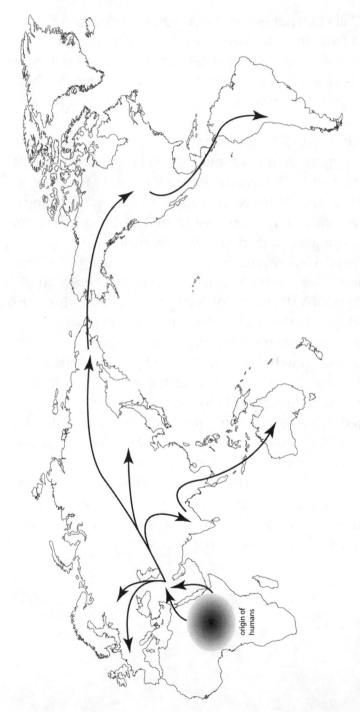

Figure 7.1. Generalized pattern of major ancient human migrations based on studies of mitochondrial and Y-chromosome DNA diversity. A single arrow may represent more than one migration.

NOTES

1. Z. A. Medvedev, *The Rise and Fall of T. D. Lysenko*, trans. I. M. Lerner (New York: Columbia University Press, 1969), pp. 11–12.

2. Ibid., p. 17.

3. Ibid., p. 58.

4. The Vavilov Institute's Web site is http://www.vir.nw.ru.

5. R. L. Cann, M. Stoneking, and A. C. Wilson, "Mitochondrial DNA and Human Evolution," *Nature* 325 (1987): 31–36; S. Horai and K. Hayasaka, "Intraspecific Nucleotide Differences in the Major Noncoding Region of Human Mitochondrial DNA," *American Journal of Human Genetics* 46 (1990): 828–42; L. Vigilant et al., "African Populations and the Evolution of Human Mitochondrial DNA," *Science* 253 (1991): 1503–1507.

6. M. Ingman et al., "Mitochondrial Genome Variation and the Origin of Modern Humans," *Nature* 408 (2000): 708–13.

7. M. F. Hammer, "A Recent Common Ancestry for Human Y Chromosomes," *Nature* 378 (2000): 376–78; M. F. Hammer et al., "Hierarchical Patterns of Global Human Y-chromosome Diversity," *Molecular Biology and Evolution* 18 (2001): 1189–1203; M. F. Hammer et al., "Human Population Structure and Its Effects on Sampling Y Chromosome Sequence Variation," *Genetics* 164 (2003): 1495–1509; R. Thomson et al., "Recent Common Ancestry of Human Y Chromosomes: Evidence from DNA Sequence Data," *Proceedings of the National Academy of Sciences, USA* 97 (2000): 7360–65; P. A. Underhill et al., "The Phylogeography of Y Chromosome Binary Haplotypes and the Origins of Modern Human Populations," *Annals of Human Genetics* 65 (2001): 43–62.

8. Hammer et al., "Human Population Structure."

9. T. Zerjal et al., "The Genetic Legacy of the Mongols," *American Journal of Human Genetics* 72 (2003): 717–21.

10. Hammer et al., "Heterogeneous Patterns of Variation among Multiple Human X-linked Loci: The Possible Role of Diversity-Reducing Selection in Non-Africans," *Genetics* 167 (2004): 1841–53.

11. L. L. Cavalli-Sforza and F. Cavalli-Sforza, *The Great Human Diasporas: The History of Diversity and Evolution* (Reading, MA: Addison-Wesley Publishing, 1995), p. 106.

12. Ibid., p. 119.

13. This note lists only a few studies because the number of studies in this area is very large. Those who wish to explore more of the original research in this area should consult the articles cited in the reference sections of the following studies: L. Quintana-Murci et al., "Where West Meets East: The Complex mtDNA Landscape of the Southwest and Central Asian Corridor," *American Journal of Human Genetics* 74 (2004): 827–45; D. Comas et al., "Admixture, Migrations, and Dispersals in Central Asia: Evidence from Maternal DNA Lineages," *European Journal of Human Genetics* 12 (2004): 495–504; N. Maca-Meyer et al., "Major Genomic Mitochondrial Lineages Delineate Early Human Expansions," *BMC Genetics* 2 (2001): 13, http://www.biomedcentral.com/1471-2156/2/13; Y. S. Chen et al., "Analysis of mtDNA Variation in African Populations Reveals the Most Ancient of All Human Continent-Specific Haplogroups," *American Journal of Human Genetics* 57 (1995): 133–49; V. Macaulay et al., "Single, Rapid Coastal Settlement of Asia Revealed by Analysis of Complete Mitochondrial Genomes," *Science* 308 (2005): 1034–36; P. Forster, "Ice Ages and the Mitochondrial DNA Chronology of Human Dispersals: A Review," *Philosophical Transactions of the Royal Society of London. Series B, Biological Sciences* 359 (2004): 255–64.

14. This note lists only a few studies because the number of studies in this area is very large. Those who wish to explore more of the original research in this area should consult the articles cited in the reference sections of the following studies: V. A. Gvozdev et al., "The Y Chromosome as a Target for Acquired and Amplified Genetic Material in Evolution," *Bioessays* 27 (2005): 1256–62; D. Charlesworth, B. Charlesworth, and G. Marais, "Steps in the Evolution of Heteromorphic Sex Chromosomes," *Heredity* 95 (2005): 118–28; M. Seielstad et al., "A Novel Y-chromosome Variant Puts an Upper Limit on the Timing of First Entry into the Americas," *American Journal of Human Genetics* 73 (2003): 700–5; A. J. Redd et al., "Gene Flow from the Indian Subcontinent to Australia: Evidence from the Y Chromosome," *Current Biology* 12 (2002): 673–77; B. Charlesworth, "The Organization and Evolution of the Human Y Chromosome," *Genomic Biology* 4 (2003): 226; T. Zerjal et al., "A Genetic Landscape Reshaped by Recent Events:

Y-chromosomal Insights into Central Asia," *American Journal of Human Genetics* 71 (2002): 466–82; R. S. Wells et al., "The Eurasian Heartland: A Continental Perspective on Y-chromosome Diversity," *Proceedings of the National Academy of Sciences, USA* 98 (2001): 10244–49; T. Karafet et al., "Paternal Population History of East Asia: Sources, Patterns, and Microevolutionary Processes," *American Journal of Human Genetics* 69 (2001): 615–28.

15. S. Wells, *Deep Ancestry: Inside the Genographic Project* (Washington, DC: National Geographic Society, 2006).

Chapter 8

THE TREE OF LIFE

When Darwin wrote the *Origin of Species*, he drew a logical, and at the time, startling conclusion. If related species are descendants of a common ancestral species, and if more distantly related species are the descendants of a more distant common ancestor, how far back could the relationships go? Might all of life ultimately trace its origin to a single source? Could humans be genetically related to everything else that is now alive or has ever lived?

Darwin included only one illustration in his book, one that depicted the relationships of living things as a tree (figure 8.1). The base of the tree represents a common ancestral species that gave rise to a group of related species. The distance from the tree's base to the tips of its branches represents time. Each branch point depicts the divergence of different types from a common ancestral species. Darwin's tree also accounts for the pattern of extinction evident in the fossil record. Not all of the twigs in his tree reach the top; those that don't represent species that have gone extinct.

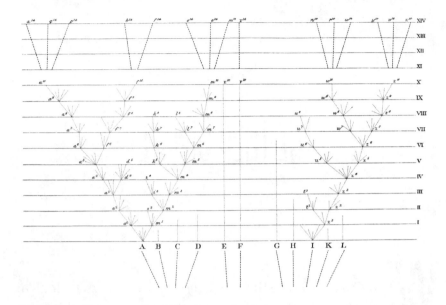

Figure 8.1. Darwin's evolutionary tree depicting the divergence of species and extinctions (from the 1876 sixth edition of *Origin of Species*).

Long before DNA evidence was available, scientists classified organisms into a great tree of life. The earliest branches separated the bacteria from eukaryotes; later branches separated major hierarchical groups of organisms. The animals, for example, could be divided into groups based on anatomical structures they had in common. Animals with a vertebral column fit into one group, those with a segmented exoskeleton (such as insects, spiders, mites, lobsters, crabs, and shrimp) fit into another group, and so on. These groups could then be further subdivided. The vertebrates could be divided into mammals, birds, reptiles, amphibians and fishes, and each of those groups could be further subdivided.

Visualizing such hierarchical subdivisions was not new; ancient Greeks and others had already thought about them. But representing them in a tree had a clear evolutionary implication. A tree of life suggests not only that life can be classified hierarchically but that the hierarchy itself can be explained as a set of ancestral evolutionary relationships—or, as Darwin put it, by "descent with modifications." According to the tree of life, all species on Earth are ultimately related to one another, some more closely, others more distantly, by common ancestry.

Anatomical comparisons have their limitations, however. Evolutionary trees are attempts to reconstruct *genetic* relationships among organisms. Anatomical similarities, for the most part, represent genetic relationships, but in some cases they can be misleading. As a simple example, cetaceans (whales and dolphins) are mammals, although anciently people thought they were fish. The reason for this mistake among the ancients is the obvious anatomical similarities that cetaceans have to fish. Their overall shape is much like that of a fish and their appendages are fins, not legs like most mammals have. However, on closer inspection, cetaceans share far more traits with other mammals. They don't have gills like fish; instead they breathe air with lungs. Their young suckle milk from mammary glands, a characteristic unique to mammals. The internal anatomy of whales and dolphins clearly distinguishes them from fish. Their paired forward fins, for example, contain the same general bone structure as the arm, wrist, and hand of a human. Five digits are buried within the fin, unlike the needle-like radiating bones of a fish's fin.

So why do dolphins and whales externally resemble fish? Darwin's theory of natural selection offers the most plausible answer. If large, land-dwelling animals had ventured into the water, any inherited variation that conferred an adaptive advantage in an aquatic environment would be favored. Variations, like webbing between the digits, give a water-dwelling organism an advantage. Given eons of time with continued natural selection, webbing between digits could eventually result in fins. Whales and dolphins look like fish because a fishlike structure confers adaptation to an aquatic environment. A penguin in water, with its finlike wings, structurally resembles fish, dolphins, and whales for the same reason.

As the ancient classification of whales and dolphins in a group with fish shows, however, anatomical similarity does not necessarily portray a close genetic relationship. A detailed and comprehensive study of anatomical similarities and differences can correct misclassification. However, the best way to determine genetic relationships is to examine not just anatomy but the genetic material itself: the DNA sequences of organisms. When DNA-sequencing technology became routine in the 1990s, scientists started to shed light on many of the questions that anatomical comparisons could not definitively answer.

One of these questions was the evolutionary origin of cetaceans. Given that dolphins and whales probably evolved from land-dwelling mammals, which modern land-dwelling mammal is most closely related to them? The answer, according to numerous studies based on extensive DNA analysis, is the hippopotamus.[1] It is the only living species of a once-large group of hippopotamuslike species called anthracotheres. The fossil record holds evidence of many species that made up this once-large group, along with the ankle bones of early aquatic whales. The branch point when the cetaceans split from anthracotheres, according to a combination of DNA analysis and fossil evidence, is fifty to sixty million years ago. Broader analysis places both the cetaceans and the anthracotheres in a larger group called the artiodactyla (the even-toed ungulates), which includes cattle, pigs, deer, camels, llamas, giraffes, antelopes, and many other cloven-hoof mammals.

DNA ANALYSIS CAN INDEPENDENTLY TEST
TRADITIONAL EVOLUTIONARY RELATIONSHIPS.

Long before DNA analysis was available, scientists developed mathematical methods for constructing evolutionary trees based on anatomical comparisons of living and extinct fossilized species. Many of the trees were published in scientific journals, and a few were illustrated in biology textbooks and popular magazines. Molecular analysis became an independent way to check the evolutionary relationships determined on the basis of anatomy. The first molecular studies were based on proteins because, for a time, the methods for comparing proteins were better than those for DNA. However, protein comparisons were expensive, time-consuming, and laborious. Only a few proteins from a relatively small sample of species were fully compared. Nonetheless, nearly all of them confirmed the traditional classifications.

One of the benefits of an enormous and costly scientific project, like sending astronauts to the moon, is the development of technologies that benefit us in ways not imagined when the project was conceived. The economic impacts of such side benefits often more than offset the cost of the project. The Human Genome Project would not have succeeded without major advances in DNA sequencing. Much of the early investment in the project focused on the development of automated, high-throughput DNA sequencing, which became rapid, inexpensive, and routine. The technology's potential is now being realized in medicine, industry, agriculture, and forensic science. Together with computer technology, it is helping evolutionary biologists to conduct large-scale comparisons of DNA sequences among many species. Staggering amounts of DNA sequence information are now being efficiently managed and statistically analyzed.

Let's use a highly simplified example to illustrate how

DNA-sequence comparisons can answer questions about evolutionary relationships. Bear in mind that this example involves only a single gene, and most studies now sample large amounts of DNA, often multiple genes, transposable elements, pseudogenes, or other types of sequences. We'll focus on the gene that encodes growth hormone in seven vertebrates: human, chimpanzee, finback whale, common dolphin, hippopotamus, camel, and alpaca (a type of llama from South America). The National Center for Biotechnology Information has posted the DNA sequences of this gene on the Internet (http://www .ncbi.nlm.nih.gov) in the centralized DNA database for the United States government. The following table lists the percentage of similarity when the growth hormone gene sequences are aligned in all possible pair-wise comparisons:

	chimp	whale	dolphin	hippo	camel	alpaca
human	0.98	0.79	0.79	0.79	0.79	0.80
chimp		0.79	0.79	0.79	0.79	0.79
whale			0.99	0.97	0.96	0.96
dolphin				0.97	0.96	0.96
hippo					0.95	0.94
camel						0.99

To use this information to construct a tree diagram, we first need to identify the pairs that match most closely. The closest are the whale-dolphin pair and the camel-alpaca pair with 99 percent similarity. The next closest, the human-chimp pair, is 98 percent similar. We can now depict the relationships by drawing lines that connect species with their closest relatives.

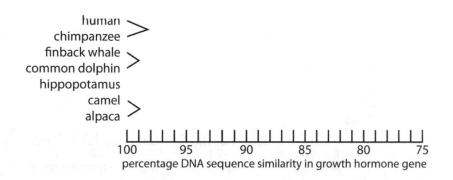

Now we look for the individual or pair that is the most similar to another individual or pair. In this case, the whale-dolphin pair is most similar to hippo; whale is 97 percent similar to hippo and so is dolphin. We can now connect the whale-dolphin pair to hippo with a branch point at 97 percent.

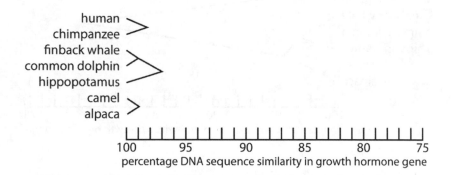

We now have three groups and need next to identify which two of the three are most similar to each other. In this case, the whale-dolphin-hippo group is most similar to the camel-alpaca group. To find out how similar, we take all pair-wise comparisons between these two groups and average them, in this case whale-camel, whale-alpaca, dolphin-camel, dolphin-alpaca, hippo-camel, hippo-alpaca. The average is 95.7 percent, so we can connect these two groups with a branch point at 95.7 percent.

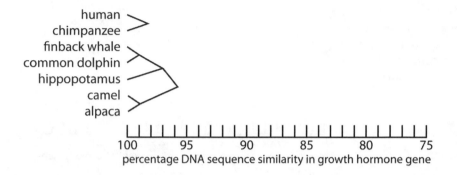

percentage DNA sequence similarity in growth hormone gene

In a final step, we connect the two remaining groups by averaging all pair-wise comparisons between the two groups. The average is 79.3 percent:

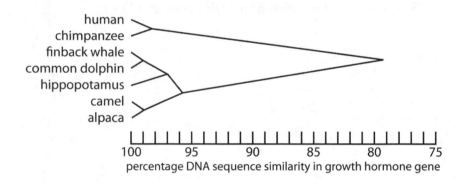

percentage DNA sequence similarity in growth hormone gene

This tree is based on DNA sequences alone, independent of any other information, yet it corroborates traditional classifications. Humans and chimpanzees are both primates and are grouped together. Whales and dolphins are cetaceans and they, too, are grouped together. The anatomical similarities of camels and llamas led scientists long ago to classify them as related species in a group called camelids, and the DNA comparison confirms that grouping. It also places whales and dolphins as most similar to the hippopotamus, and that group is most similar to the camel-alpaca group, which consists of even-toed ungulates. The close relationship of cetaceans with hip-popotamus, and more broadly with even-toed ungulates, has been confirmed by multiple comparisons like this one, although much more elaborate.

The method for tree construction we examined here is highly simplified. More-advanced methods employ sophisticated mathematical analyses and computer algorithms to reconstruct far more informative and complex evolutionary relationships. They offer many advantages, including identification of which variations are ancestral and which are more recent mutations. Retroelements and pseudogenes are especially valuable for this type of evolutionary analysis because they represent one-time insertion events. They are not excised precisely from their insertion sites once they are there, so a sequence without a retro-element or pseudogene must be the ancestral type when compared to the same sequence in another species that does have such an insertion. Appendix 1 has more detail on this topic.

DNA SEQUENCES ARE BEING USED TO CONSTRUCT A TREE OF LIFE FROM THE DEEPEST ROOTS TO THE FINEST BRANCHES.

High-throughput automated DNA sequencing is being used to develop a highly accurate tree of life. The National Science Foundation (NSF) of the United States has embarked on a long-term project of assembling the tree of life, funding research at many universities where different branches of the tree are being constructed. According to the NSF, "Resolving the Tree of Life is unquestionably among the most complex scientific problems facing biology and presents challenges much greater than sequencing the human genome."[2]

Assembling the tree of life is of enormous value to science because it offers the clearest and most accurate understanding of how living species are related to one another. It also has many practical advantages as the information it generates is applied to medicine, agriculture, industry, and wildlife management. The challenge, however, is great. So far, about 1.75 million species have been scientifically documented, an almost incomprehensible number. However, this represents just a small fraction of the number of species currently on Earth, which by most estimates are in the tens of millions.[3] Those species are the twigs on the tree of life, which is both enormous and enormously complex.

At this point in the book, three observations should be apparent. First, the results of hundreds of large-scale experiments based on DNA analysis overwhelmingly confirm the reality of evolution. Second, the vast amount of DNA sequence information now available allows scientists to accurately reconstruct evolutionary relationships among species currently alive on Earth. Third, conclusions derived from DNA analysis can be integrated with information from traditional studies of anatomy, physiology, the fossil record, archaeology, and geology to decipher the history of life on Earth.

More than one hundred and fifty years of investigations by thousands of scientists have yielded vast and powerful evidence of evolution. Yet some people still claim that evolution is a "theory in crisis,"[4] unsupported by solid evidence, and one that should be abandoned. The next chapter addresses the ongoing assault on science by highly organized and well-funded groups whose political objectives are to cast doubt on the reality of evolution and to restrict or dilute it in the science curricula of public schools.

NOTES

1. D. Graur and D. G. Higgins, "Molecular Evidence for the Inclusion of Cetaceans within the Order Artiodactyla," *Molecular Biology*

and Evolution 11 (1994): 357–64; J. Gatesy et al., "Evidence from Milk Casein Genes that Cetaceans Are Close Relatives of Hippopotamid Artiodactyls," *Molecular Biology and Evolution* 13 (1996): 954–63; M. Shimamura et al., "Molecular Evidence from Retroposons that Whales Form a Clade within Even-Toed Ungulates," *Nature* 388 (1997): 666–70; R. G. Kleineidam et al., "Inclusion of Cetaceans within the Order Artiodactyla Based on Phylogenetic Analysis of Pancreatic Ribonuclease Genes," *Journal of Molecular Evolution* 48 (1999): 360–68; M. Nikaido, A. P. Rooney, and N. Okada, "Phylogenetic Relationships among Cetartiodactyls Based on Insertions of Short and Long Inter-spersed Elements: Hippopotamuses Are the Closest Extant Relatives of Whales," *Proceedings of the National Academy of Sciences, USA* 96 (1999): 10261–66; U. Arnason et al., "The Mitochondrial Genome of the Sperm Whale and a New Molecular Reference for Estimating Eutherian Divergence Dates," *Journal of Molecular Evolution* 50 (2000): 569–78; Z. Maniou, O. C. Wallis, and M. Wallis, "Episodic Molecular Evolution of Pituitary Growth Hormone in Cetartiodactyla," *Journal of Molecular Evolution* 58 (2004): 743–53.

2. From a pamphlet on the NSF tree of life available as a pdf file at http://www.ucjeps.berkeley.edu/tol.pdf (accessed May 25, 2005).

3. Ibid.

4. This phrase is derived from the title of the book: M. Denton, *Evolution: A Theory in Crisis* (Bethesda, MD: Adler and Adler, 1986).

Chapter 9

"WHEN FAITH AND REASON CLASH"[1]

Most of us look upon nature with wonder, filled with admiration for its variety and beauty. Some feel that science snuffs out the wonder of nature by unraveling its mysteries, as if somehow the mystique of the unknown lends nature a sense of awe. To others, science magnifies nature's wonder by revealing unimaginable intricacies, treasures that for all of human history have remained hidden until now. This book has highlighted just a few of those previously hidden treasures.

Throughout the ages, many people have looked upon the marvels of nature and found comfort in the idea that a God with infinite intelligence must have created them. To many, the complexity of nature appeared too great to conclude otherwise. The first chapter of the Bible seemed to explain why there was so much complexity. God had apparently created each species separately, and these species had been preserved essentially unchanged to the present day. A literal interpretation of the book of Genesis presumes an Earth that is about six thousand years old, with all species created at its beginning. Then a selected number of them were preserved in a single vessel of known dimensions during a worldwide flood to repopulate the earth after life upon it was destroyed.

Ever since the 1700s, scientific inquiries, particularly in biology and geology, have been turning up evidence that runs counter to a literal biblical interpretation of Earth's natural history. The accumulating evidence pointed to a world that was billions of years old, populated by successive arrays of species that appeared related to one another in a hierarchical fashion. The number of cataloged species became far too great and their distributions throughout the earth far too complex to be explained by a single recent creation and preservation of all current terrestrial species in an ark during a worldwide flood. The conflict between science and traditional religious interpretations became especially heated in the 1800s as a number of scientists proposed evolutionary origins for living things. Charles Lyell's evidence of a far more ancient Earth had also gained prominence. To many, the notion of a six-thousand-year-old Earth and the doctrine of special creation were becoming increasingly less plausible.

The publication of *Origin of Species* in 1859 set off a firestorm. Scientists throughout the world accepted that life had evolved because evolution so clearly explained a host of phenomena in nature. Some, but not all, members of the clergy perceived evolution as an assault on Christianity and the Bible. Atheists seized the evidence of evolution as a sword to attack religion as outdated myth, no longer acceptable in enlightened societies. Religious zealots embraced the opportunity to put on God's armor and passionately defend the faith.

Darwin's views on religion were ambiguous and changed during the course of his life. His private writings reveal his dissatisfaction with many of the doctrines of nineteenth-century Christianity. Publicly, however, he sought to quench any controversy between his theory and religion. He was a kind and gentle family man, and the controversy sparked by his theory troubled him.

He knew that his theory was incompatible with the doctrine of special creation and would generate conflict. Ever the optimist, he hoped that as scientists pursued knowledge about nature, his concept of natural selection would win out in the end. In his words from the *Origin of Species*: "It is so easy to hide our ignorance under such expressions as the 'plan of creation,' 'unity of design,' &c. . . . A few naturalists, endowed with much flexibility of mind, and who have already begun to doubt the immutability of species, may be influenced by this volume; but I look with confidence to the future, to young and rising naturalists, who will be able to view both sides of the question with impartiality."[2] His hope would be fulfilled well beyond his dreams; among modern biologists, evolution and the explanatory power of natural selection have near-unanimous acceptance.

Darwin also felt compelled to comment on religion. In the concluding chapter of the *Origin of Species* he wrote,

> I see no good reason why the views given in this volume should shock the religious feelings of any one. It is satisfactory, as showing how transient such impressions are, to remember that the greatest discovery ever made by man, namely, the law of the attraction of gravity, was also attacked by Leibnitz, "as subversive of natural, and inferentially of revealed, religion." A celebrated author and divine has written to me that "he had gradually learnt to see that it is just as noble a conception of the Deity to believe that He created a few original forms capable of self-development into other and needful forms, as to believe that He required a fresh act of creation to supply the voids caused by the action of His laws."[3]

Darwin's book concludes with a reference to the Creator in perhaps its most poetic passage: "There is grandeur in this view of life, with its several powers, having been breathed by the Creator into a few forms or into one; and that whilst this planet has gone cycling on according to the fixed law of gravity, from so simple a beginning endless forms most beautiful and most wonderful have been, and are being, evolved."[4] It is particularly telling that the words "by the Creator" were not in this passage in the first edition. Darwin added them in the second edition and retained them through all subsequent editions.

Attempts by Darwin and others to reconcile Christianity and evolution were deemed unacceptable by many people, especially in the United States. Fundamentalist Christian preachers taught that the two were completely incompatible; to accept one was to reject the other. The clash between evolution and religion grew heated in Darwin's day. Since then it has ebbed and flowed but never subsided.

During the early twentieth century, the scientific study of life—biology—expanded rapidly from an infant science into a complex set of subdisciplines. Discoveries began to produce astounding benefits for humanity. They transformed medicine, expanding the arsenals against disease and extending the average life span. Application of genetic and evolutionary principles to agriculture multiplied world food production several-fold. Petroleum and coal, the compacted and chemically transformed remnants of long-extinct plants, were fueling the industrial revolution. Biology, and its companion sciences of chemistry, physics, and geology, truly were revolutionizing the world. The rediscovery of Gregor Mendel's principles of inheritance in 1900 and the rapid progression of genetics in subsequent decades filled a major gap in Darwin's theory: the role of inheritance in evolution. The theory took on a distinctly modern tone when merged with genetics, and evolution soon became the unifying theme of modern biology. The 1973 title of an article by the famous evolutionary geneticist Theodosius Dobzhansky said it all: "Nothing in Biology Makes Sense Except in the Light of Evolution."[5]

During the early twentieth century, as evolution increasingly unified biology, the debate over its implications for religion intensified. Nowhere has the debate remained more intense than in discussions of how evolution should be taught (or not taught) in public schools in the United States. By the 1920s, several state legislatures perceived evolution as a threat to faith and legally banned it from the public schools. These laws became popularly known as "monkey laws" because they specifically forbade the idea that humans were a product of evolution or had common ancestry with apes and monkeys. They set the stage for one of the best-known and most dramatic court cases in all of legal history.

THE SCOPES TRIAL ATTRACTED NATIONAL ATTENTION TO THE CREATION-EVOLUTION CONTROVERSY.[6]

The most famous legal battle over religion and evolution was the 1925 Scopes trial held in Dayton, Tennessee. That year, the Tennessee state legislature had passed House Bill 185, which prohibited "the teaching of the Evolution Theory in all the Universities, Normals and all other public schools of Tennessee, which are supported in whole or in part by the public school funds of the State. . . ." It specifically targeted human evolution. Teachers were forbidden "to teach any theory that denies the story of the Divine Creation of man as taught in the Bible, and to teach instead that man has descended from a lower order of animals." The American Civil Liberties Union (ACLU) sought for an opportunity to challenge the Tennessee law in court. A group of prominent Dayton citizens saw such a trial as a chance to publicize their town and bring business to it. John Scopes, a high school general science teacher, had been substituting for the regular biology teacher and had assigned readings from a biology textbook that explained evolution. Officers interrupted Scopes while he was playing tennis to ask if he would agree to stand trial, and he concurred. He was officially charged with violating the law; he then promptly returned to his tennis game.

The Scopes trial turned out to be a heated battle between two nationally famous attorneys. The prosecuting attorney was William Jennings Bryan, who had used his vehement opposition to evolution as a rallying cry in his three (unsuccessful) presidential campaigns. Clarence Darrow, a skilled litigator and agnostic orator, agreed to defend Scopes. The national attention generated by the trial exceeded the hopes of Dayton's promoters. The town's atmosphere was like a circus; it made daily headlines nationwide and was the first trial to be carried live by radio.

The trial itself turned into the spectacle everyone had hoped it would be. The prosecution presented a brief case, documenting through testimony from the school superintendent and several students that Scopes had indeed taught evolution, including human evolution, and thus had violated the law. After the prosecution rested, the defense called its first witness, Dr. Maynard Metcalf of Johns Hopkins University, as an expert on evolution. The prosecution objected, claiming that expert testimony on evolution was irrelevant to the proceedings. After listening to part of Dr. Metcalf's testimony, and to impassioned arguments from both the prosecution and the defense, the judge excluded expert testimony on evolution from the trial.

This ruling evoked an angered response from Darrow, who had arranged for several renowned scientists and theologians to testify. His outburst earned him a contempt citation, but the judge revoked the citation when Darrow later apologized. In what turned out to be a brilliant maneuver that made legal history, Darrow called the prosecuting attorney, William Jennings Bryan, as a witness for the defense and as an expert on religion. Bryan agreed to testify, in spite of protests from his assistants, seeing this as a chance to promote his cause to a national audience. Darrow's strategy succeeded and Bryan's backfired as Darrow skillfully humiliated Bryan with recurring questions on whether or not the Bible should be interpreted literally. At one point, in frustration, Bryan exclaimed, "I do not think about things I don't think about." Darrow retorted, "Do you think about things you do think about?" The exchange between Darrow and Bryan grew increasingly adversarial until the judge finally put a stop to the questioning. The following day, the judge ruled that Bryan could no longer testify and that his previous testimony should be stricken from the record. By then, Darrow had won the war of public opinion but ended up losing the case when the jury convicted Scopes.

In reality, Darrow expected to lose the case and hoped to appeal it, preferably all the way to the United States Supreme Court. He hoped that the US Supreme Court would nullify the Tennessee statute and, along with it, all other antievolution laws. The ACLU first appealed the case to the Tennessee Supreme Court, which reaffirmed the law but dismissed the case on a technicality. This ruling exonerated Scopes, preventing further appeals and dealing a blow to opponents of antievolution laws who had hoped that the US Supreme Court would eventually hear the case. The law remained in force, as did similar laws in other states.

THE AFTERMATH OF THE SCOPES TRIAL
HAD SERIOUS REPERCUSSIONS FOR SCIENCE EDUCATION.

The outcome of the Scopes trial was a mixed bag for both sides. Most newspaper and radio reporters cast the opponents of evolution as uneducated and backward. Nationally, public opinion seemed to favor the supporters of evolution. However, what the supporters gained in public opinion, they lost in public schools. Fearing reprisals from school districts and parents, textbook companies carefully excluded evolution from biology textbooks and teachers often avoided teaching it in their classes. At a time when evolution was increasingly perceived as the central theme of modern biology, it was largely absent from grade school and high school biology. This situation persisted for more than three decades.

When the Soviet Union successfully placed the *Sputnik* satellite in orbit in 1957, the US government panicked over the perceived crisis in American science. In response, the National Science Foundation initiated a review of science curricula in public schools. A group called the Biological Sciences Curriculum Study (BSCS) studied biology teaching and found courses and textbooks to be inexcusably deficient. The group developed a set of modern biology textbooks that highlighted the centrality of evolution in modern biology. The first set of textbooks was published in 1963 and was rapidly adopted by schools throughout the nation. Private textbook publishers followed suit. In short order, evolution was vaulted from near exclusion to standard fare in biology classrooms in public schools.

The legal tide was turning as well. The 1925 Tennessee law that had become famous in the Scopes trial was still in force, but the Tennessee legislature repealed it in 1967—forty-two years after its passage. The following year, the first antievolution law to reach the US Supreme Court was declared unconstitutional by the Court in *Epperson v. Arkansas*. By then, most state legislatures had repealed laws prohibiting the teaching of evolution; only Arkansas and Mississippi still had such laws in force and both were nullified by the Supreme Court's ruling. Evolution, it seemed, had gained a strong foothold in American science education.

THE CREATION-SCIENCE MOVEMENT
ATTEMPTS TO CAST CREATIONISM AS SCIENCE.

Christian fundamentalists feared that the increasing acceptance of evolution would undermine faith, particularly among children as they learned about evolution in school. Henry M. Morris, a hydraulic engineer who was a distinguished professor at three major universities, rapidly established himself as the leader and foremost spokesman for a movement known alternatively as creationism and creation science. The movement adopted biblical literalism and used it to vehemently oppose evolution. It claimed a creation of six twenty-four-hour days, fixity of "created kinds" of animals and plants, separate and special creation of humans, and a worldwide flood that destroyed all terrestrial animals and humans except those preserved on Noah's ark.

As a scientist (albeit not a biologist or geologist), Morris attempted to build a scientific basis for biblical literalism. His most significant work, which he coauthored with theologian John Whitcomb, was a book titled *Genesis Flood*.[7] This book attempted to provide a scientific basis for a young universe (no older than ten thousand years) created out of nothing, and a worldwide flood sufficiently large enough to cover the whole of the earth, including the tops of the tallest mountains. According to *Genesis Flood*, the fossil record consisted of the plants and animals killed during the flood and the sequential deposition of their bodies in sediments generated by the cataclysmic flooding over a period of days, not the hundreds of millions of years proposed by modern science. The book further claimed that pairs of created kinds of animals preserved on Noah's ark were the original parents of all terrestrial animals that repopulated the post-flood Earth. Today, after more than forty years of success, *Genesis Flood* remains in print as a creationist classic with widespread sales and readership.

In an effort to promote so-called creation science, Morris joined with others to form the Creation Research Society in 1963. Partly in response to the reintroduction of standard scientific evolution to public schools and textbooks, the repeal of antievolution laws, and the Supreme Court ruling against such laws, creationists established their own college, called the Christian Heritage College in 1970. The college's research division was renamed the Institute for Creation Research (ICR) in 1972 and became autonomous from the college in 1981. Over the years, the ICR grew into the leading creationist organization in the United States. It continues to sponsor books and other publications, research, seminars, conferences, workshops, debates, a graduate school in creation science, and a museum devoted to creation science.[8]

The ICR also attempted to promote new legislation requiring that any time "evolution science" was taught, equal time had to be devoted to "creation science." Success was mixed. Equal-time legislation was introduced in twenty-seven states but succeeded only in Arkansas and Louisiana.[9] The Arkansas law became the first legal test case. Known as Act 590, passed in 1981, it set the standard for legally defining creation science. US District Court Judge William R. Overton heard the challenge to the law and ruled that the law was unconstitutional. The essential reason was that creation science is religion, not science, so it violated the establishment clause of the Constitution, which mandates separation of church and state.

Overton's ruling was significant because it clearly defines the realm of science. It is based on testimonies of leading scientists and science philosophers and set a precedent for subsequent cases. In reference to the creation-science claim of ex nihilo creation (a sudden creation out of nothing), Overton wrote, "Such a concept is not science because it depends upon a supernatural intervention which is not guided by natural law."[10] Perhaps the most telling conclusion had to do with the relationship of evolution to religion: "The theory of evolution assumes the existence of life and is directed to an explanation of *how* life evolved. Evolution does not presuppose the absence of a creator or God and the plain inference conveyed by Section 4 [that it does] is erroneous."[11]

These conclusions are not Overton's alone, and they accurately represent the majority of scientists worldwide, regardless of whether they are religious. The key component is the absolute requirement that science be based on natural law and that its conclusions be empirically testable. The supernatural, by definition, is "beyond nature." It falls outside scientific explanations of the natural world, which are based on observations and experimental tests that yield consistent results when repeated.

Science does not exclude the existence of God as Creator or the possibility of supernatural intervention. However, while not denying the possibility of their existence, it cannot include or infer supernatural acts or intervention because neither can be subjected to experimental tests in the real world. By definition, "creation science" is not science.

Louisiana was the only other state to pass an equal-time law, which likewise was challenged in court. As in the Arkansas case, the judge determined that creation science was religion, not science, and declared the law unconstitutional. The Arkansas ruling was never appealed, but the Louisiana one was, and the appellate court upheld the ruling. The case made its way to the US Supreme Court, which in 1987 upheld the lower-court rulings.

Beyond the legal implications, most scientists find the claims of the creation-science movement to be completely at odds with the overwhelming body of scientific evidence. For example, Kenneth Miller, a cell biologist and professor at Brown University, pointed out in his book *Finding Darwin's God*[12] that the claim that the earth and the entire universe were created out of nothing less than ten thousand years ago runs counter to abundant geological, radiometric, and astronomical evidence from multiple independent studies.[13] According to Miller, if God created the universe only ten thousand years ago, then the billions of stars more than ten thousand light-years away should be invisible to us because their light has not yet reached us.

Creationist authors recognize the serious scientific problems a ten-thousand-year-old creation invokes, but they explain them away in a strange, scientifically untestable way. They say that the Creator intentionally gave Earth and the entire universe the "appearance of age." Whitcomb and Morris repeatedly appeal to this explanation in *Genesis Flood*. Regarding the problem with light from the stars reaching Earth, they wrote:

> A common opinion is that the very distance of the far galaxies testifies that the universe must be billions of years old. Since these galaxies are known to be some few billion light-years away, by definition it has taken that number of years for their light to reach us; therefore they are at least that old, so the argument goes.
>
> But this contention of course again begs the question. It constitutes an implicit denial that the universe could have been created as a functioning entity. . . . It must have had an "appearance of age" at the moment of creation. The photons of light energy were created at the same instant as the stars from which they were apparently derived, so that an observer on earth would have been able to see the most distant stars within his vision at that instant of creation. There is nothing unreasonable either philosophically or scientifically in this . . .[14]

Whitcomb and Morris also make similar claims about the "appearance of age" in the results of radiometric dating of certain minerals, which tell us that Earth and the rest of our solar system is about 4.5 billion years old. God, they decided, made the solar system appear to be that old when, in fact, it is much younger.

Miller minces no words in pointing out the absurdity, even the danger, of such claims:

> In order to defend God against the challenge from evolution, they [the creationists] have had to make Him into a schemer, a trickster, even a charlatan. Their version of God is one who intentionally plants misleading clues beneath our feet and in the heavens themselves. Their version of God is one who has

filled the universe with so much bogus evidence that the tools of science can give us nothing more than a phony version of reality. In other words, their God has negated science by rigging the universe with fiction and deception. To embrace that God we must reject science and worship deception itself.[15]

Among the most ironic claims of the creationists is the notion that genetic changes are possible "only within fixed limits of originally created kinds."[16] They conspicuously avoid the word *species*, opting instead for the biblical word *kind*. They do this to reconcile a glaring problem with their claim of a literal biblical interpretation that a worldwide flood destroyed all terrestrial animal life. The book of Genesis specifies the dimensions of Noah's ark, and it was far too small to accommodate pairs of even a fraction of the known land-dwelling species of animals. The Scopes-era creationist, George McCready Price, came up with an explanation that was later embellished by many creationist authors, including Whitcomb and Morris.[17] It invokes a strange type of evolution (currently called "rapid post-Flood adaptation" to avoid the word *evolution*) in which multiple species arose by way of natural selection from an original "created kind." They see the so-called created kinds as major groups of organisms, each represented by one pair on the ark.

As creationist Carl Wieland put it in an article published in *Creation*, a journal for creationist authors: "Creationists have long proposed such 'splitting under selection' from the original kinds, explaining for example wolves, coyotes, dingoes and other wild dogs from one pair on the Ark."[18] In other words, new species supposedly evolved during the past 5,300 years by a tremendously accelerated version of Darwinian natural selection from common ancestral kinds on the ark, but each kind was specially created and could not change into another kind.

Defining what a "kind" is has been problematic for creationists, although they all agree on one thing: humans are a distinctly separate, specially created kind, and therefore do not share common ancestry with any other animal, including other primates. The rate at which new species purportedly evolved from original kinds after the flood is *far* faster than the rates proposed by evolutionary biologists for the divergence of these same species from ancient common ancestry. They claim it took only a few thousand years at most to produce the vast diversity of terrestrial animal species instead of tens to hundreds of millions of years as determined by scientific studies. Not surprisingly, this pseudo-Darwinian claim of "rapid post-Flood adaptation" dismays biologists and antievolutionists alike.

Miller aptly summed up his analysis of creationist claims when he wrote, "Such so-called creation science, thoroughly analyzed, corrupts both science and religion, and it deserves a place in the intellectual wastebasket."[19] In spite of such vociferous and nearly unanimous rejection of creation science by scientists, the Institute for Creation Research continues to thrive, supported by a large number of people who count themselves as biblical literalists. As shown by Gallup polls, 45 percent of Americans accept the creationist view that God created humans in a single act less than ten thousand years ago.[20]

THE INTELLIGENT DESIGN MOVEMENT CLAIMS THAT THE COMPLEXITY OF LIVING ORGANISMS IS THE WORK OF AN INTELLIGENT DESIGNER.

The latest challenge to evolution comes from a movement known as "intelligent design." The movement is a modern revival of an old argument articulated by the eighteenth-century naturalist and theologian William Paley. In his greatest work, *Natural Theology*, Paley argued that one can readily distinguish something that has been designed from something that is not:

In crossing a heath, suppose I pitched my foot against a *stone* and were asked how the stone came to be there, I might possibly answer that for anything I knew to the contrary it had lain there forever; nor would it, perhaps, be very easy to show the absurdity of this answer. But suppose I had found a *watch* upon the ground, and it should be inquired how the watch happened to be in that place, I should hardly think of the answer which I had before given, that for anything I knew the watch might have always been there. Yet why should not this answer serve for the watch as well as for the stone? Why is it not as admissible in the second case as in the first? For this reason, and for no other, namely, that when we come to inspect the watch, we perceive—what we could not discover in the stone—that its several parts are framed and put together for a purpose, e.g., that they are so formed and adjusted as to produce motion, and that motion so regulated as to point out the hour of the day; that if the different parts had been differently shaped from what they are, of a different size from what they are, or placed after any other manner or in any other order than that in which they are placed, either no motion at all would have been carried on in the machine, or none which would have answered the use that is now served by it.[21]

The evidence of design in a watch, according to Paley, is the fact that it functions only when the various parts are just the right sizes and shapes to work together so that the watch carries out its useful motion. Moreover, if any part of the watch were to be significantly altered or removed, the watch would cease to function. The very existence of a watch points to a watchmaker who is an intelligent designer. The watch, says Paley, is analogous to nature, which, being so much more intricate and complex than a watch, must have been created by an intelligent designer, and the designer is God.

The intelligent design movement casts Paley's argument in a modern context. Unlike creationism, which relies on biblical literalism, intelligent design advocates rarely discuss the Bible in their writings. Instead, they focus on two major strategies. First, they claim the present range of diversity and complexity of living organisms is too great to be explained by undirected Darwinian natural selection, so an intelligent designer must have directed the origin and development of life. Second, they attempt to find as many flaws as possible, both real and contrived, in the writings of evolutionary biologists. The idea is to chip away at modern evolutionary theory as unfounded, conflicted, and unreliable.

Intelligent design advocates know this second strategy as the "wedge," so named by Phillip Johnson, a law professor at University of California, Berkeley. Johnson brought the intelligent design movement into prominence with his book *Darwin on Trial*.[22] Most books and articles promoting intelligent design have focused on "driving a wedge" into a perceived evolution establishment.

The wedge strategy is effective because scientists tend to scrutinize the experimental rationale and reasoning in one another's work. Modern science by its very nature favors novel approaches to experimentation and interpreting nature. It demands rigorous tests of predictions derived from hypotheses about the natural world. It requires collecting preferably quantifiable data and analysis of experimental results.

Especially when scientists are pioneering new, uncharted territories, they often disagree about how to interpret new information. Disagreement does not necessarily mean an interpretation is wrong. Rather, it means the scientific community remains open to probing interpretations in rigorous ways. Such probing helps us expand and refine our understanding of nature.

Evolutionary biology is no exception. An overwhelming majority of scientists accept that the present and past diversity of life came about by evolution.[23] However, they continue to probe the mechanisms of evolution, and they often disagree over how to interpret newly discovered details. How vociferously they disagree over an evolutionary event is roughly proportional to the time that has elapsed since that event. In general, the longer ago an evolutionary event happened, the less likely we are to have abundant and easily discernable information about it. An evolutionarily remote event supported by fragmentary data is likely to be more controversial among scientists than a recent event supported by a wealth of highly reliable data.

For example, the divergence of separate lineages leading to humans and chimpanzees from a common ancestor about six million years ago is an evolutionarily recent event (although six million years may not seem recent, it is when compared to more than two billion of years of evolution). The evolutionary relationship of humans and chimpanzees is exceptionally well supported by a large body of evidence, a fraction of which is highlighted in this book. There is practically no disagreement among scientists regarding the common ancestry of humans and chimpanzees. How multicellular organisms evolved from single-celled bacteria over a billion years ago, however, is a highly disputed subject. Most of the wedge strategy literature focuses on such disputes, often employing a tactic known as "quote mining," selecting out-of-context quotation from articles by reputable scientists in an attempt to highlight the disagreement, while excluding the overwhelming consensus about evolution and its principal mechanisms.

Scientists have not been silent in the face of the wedge strategy attack on evolution. A number of books challenge creationist and intelligent design claims.[24] One of the most complete online resources, the Talk.Origins Archive, is a mainstream science resource "devoted to the discussion and debate of biological and physical origins" (http://www.talkorigins .org). It includes detailed and well-documented rebuttals to the most popular wedge strategy books, articles, and Web sites.

For all their calls to oppose evolutionary thought, intelligent-design advocates have not come up with a scientifically plausible alternative. Their usual alternative of choice is the hypothesis of irreducible complexity proposed by Michael Behe, a professor of biochemistry at Lehigh University in Pennsylvania. In *Darwin's Black Box*,[25] he attempts to respond to a challenge issued by Darwin: "If it could be demonstrated that any complex organ existed which could not possibly have been formed by numerous, successive, slight modifications, my theory would absolutely break down."[26]

According to Behe, many complex organs and biochemical systems in living organisms could not have been formed by successive modifications because they are irreducibly complex. His definition of an irreducibly complex system is "a single system composed of several well-matched, interacting parts that contribute to the basic function, wherein the removal of any one of the parts causes the system to effectively cease functioning."[27] He argues that an irreducibly complex system cannot arise in a step-by-step fashion because each component is useless without the other components; all parts of the entire system must be formed simultaneously for the system to work: "Since natural selection can only choose systems that are already working, then if a biological system cannot be produced gradually it would have to arise as an integrated unit, in one fell swoop, for natural selection to have anything to act on."[28] Behe further claims that "as the number of unexplained, irreducibly complex biological systems increases, our confidence that Darwin's criterion of failure has been met skyrockets toward the maximum that science allows." The advocates of this view claim that if Darwinian selection cannot explain the origin of irreducibly complex systems, then, by default, we must attribute their origin to an intelligent designer.

You may have noticed that the idea of irreducible complexity is similar to Paley's argument of design with a watch as an example. According to Paley, if one part of the watch is removed, the watch ceases to function; therefore, by Behe's definition, the watch is an irreducibly complex system. Behe spends most of his book discussing examples of irreducibly complex biological systems. No biologist would disagree that biological systems are complex and that removal of certain parts of these systems reduces or eliminates their function. However, it is quite a leap to claim that irreducibly complex systems cannot arise through natural processes. Let's focus our attention on one of the examples Behe cites in his book: the vertebrate blood-clotting system.

A cut on your skin ruptures blood vessels, and you begin bleeding. Fortunately, you have a blood-clotting system that rapidly forms a mesh of protein fibers to stop the bleeding. The system is quite complex with multiple proteins (each encoded by a gene) interacting with one another to form the clot. The system is a cascade in which one protein activates another, which, in turn, activates yet another protein, and so on, magnifying the signal with each step in the cascade. If a protein in the cascade is missing, the clotting system either fails or loses some of its ability to function. Hemophilia (excessive bleeding) is a human disease caused by a mutation in one gene that eliminates the function of one protein in the cascade. According to Behe's definition, our blood-clotting system is irreducibly complex, at least for some of the components in the system.

But is evolution incapable of producing it? Claiming that evolution could not have produced it, Behe attacks the work of Russell Doolittle, a highly respected scientist at the University of California, San Diego, who has spent much of his career studying the evolution of the vertebrate blood-clotting system. Behe focuses on one of Doolittle's review articles, which explains in broad terms the most likely path by which the blood-clotting system evolved. Interestingly, most of the proteins in the cascade are similar to one another, and so are the genes that encode them. The evolutionary scenario proposed by

Doolittle is one we encountered earlier in this book in which the system grows in complexity through duplication and divergence of its genes. In the case of the genes in the blood-clotting system, there is evidence that pieces of some of the duplicated genes have been transferred to other similar genes in the system to create yet greater gene diversity.

Recognizing the similarity of the genes, Behe claims that Doolittle's scenario is mathematically so unlikely that undirected evolution could not possibly have produced it:

> Doolittle appears to have in mind a step-by-step Darwinian scenario involving the undirected, random duplication and recombination of gene pieces. But consider the enormous amount of luck needed to get the right gene pieces in the right places. Eukaryotic organisms have quite a few gene pieces, and apparently the process that switches them is random. So making a new blood-coagulation protein by shuffling is like picking a dozen sentences randomly from an encyclopedia in the hope of making a coherent paragraph. Professor Doolittle does not go to the trouble of calculating how many incorrect, inactive, useless "variously shuffled domains" would have to be discarded before obtaining a protein with, say, TPA-like activity. [TPA is one of the proteins in the blood-clotting cascade.]

To illustrate the problem, let's do our own quick calculation. Consider that animals with blood-clotting cascades have roughly 10,000 genes, each of which is divided into an average of three pieces. This gives a total of about 30,000 gene pieces. TPA has four different types of domains. By "variously shuffling," the odds of getting those four domains together is 30,000 to the fourth power, which is approximately one-tenth to the eighteenth power. Now, if the Irish Sweepstakes had odds of winning of one-tenth to the eighteenth power, and if a million people played the lottery each year, it would take an average of about a thousand billion years before *anyone* (not just a particular person) won the lottery. A thousand billion years is roughly a hundred times the current estimate of the age of the universe. . . . Doolittle apparently needs to shuffle and deal himself a number of perfect bridge hands to win the game. Unfortunately, the universe doesn't have time to wait.

On the surface, Behe's assessment might seem convincing. But let's analyze it a bit to see whether or not it is scientifically reliable. Incidentally, he has made three minor errors that require correction before we move on. First, his calculation of thirty thousand to the fourth power is mathematically incorrect; it should be one divided by thirty thousand to the fourth power. Second, genomic studies point to twenty thousand or more genes in vertebrate genomes, not ten thousand. And third, the number of "pieces" in a gene is typically more than three (Behe is referring to exons as "pieces"). These two latter corrections actually work *in favor* of Behe's argument, reducing the probability even more than he proposes.

However, Behe also makes two serious substantive errors. First, when calculating the likelihood that an event will happen, we must take into account the number of opportunities for that event to occur. Behe does this by analogy, referring to a million people playing the Irish lottery. In an actual evolutionary scenario, the number of opportunities is the number of individuals in ancestral species that can pass genes on to offspring. When we consider that the number of individuals in a single species at any one time may number in the millions, billions, or more depending on the life cycle and reproductive success, and that the species may persist for millions of years, the number of opportunities is literally incalculable but certainly far, far greater than one million. For example, suppose gene shuffling produces a new version of a gene in the vertebrate blood-clotting system. To calculate the number of opportunities of that event happening in just the vertebrates alive today, we have to count every human, every mammal, every reptile, every amphibian, and every fish alive right now on the entire planet, which, of course, is a practically impossible task, but we can safely surmise that the number is exceptionally large. Now, if we extend that back over the hundreds of millions of years that vertebrates have been around, the probability of even an extremely unlikely event happening is actually quite high. Not only does the universe have time to wait, Earth has had plenty of time and opportunities.

Behe's second substantive error renders this first error moot. His claim that Doolittle has in mind "random duplication and recombination of gene pieces" is incorrect. Gene duplication is not necessarily random—as we saw when discussing transposable elements and pseudogenes, repeated sequences in DNA predispose certain genes to duplication—and gene recombination is certainly not random. The type of gene recombination between similar genes, such as those in the blood-clotting system, is called *gene conversion*. This event happens when the DNA of two genes with similar sequences end up juxtaposed with each other. Because the sequences are similar, strands of DNA from both genes may exchange places, forming base pairs with the other gene. If the exchanged DNA remains in place, a piece of one gene is transferred to another, resulting in a recombined gene. The event is not random; it happens only between similar DNA sequences.

The similarity among many of the different genes in the blood clotting system predisposes them to recombination through gene conversion. These recombinations are very rare events, but they have been documented in many species, including humans. Interestingly, a recent example of gene conversion in humans is in the blood-clotting system,[29] the very type of gene shuffling that Behe claims is so unlikely.

The most serious flaw in the concept of irreducible complexity is Behe's claim that an irreducibly complex system "would have to arise as an integrated unit, in one fell swoop."

The human blood-clotting system currently functions in a high-pressure circulatory system. However, evolution of one system is not independent of the evolution of related systems. The high-pressure circulatory system of modern mammals apparently arose gradually from a primitive low-pressure system as the need for improved blood circulation increased in species with more complex bodies. Low-pressure circulatory systems do not require the multiple-step clotting process of a high-pressure system. As selection favored gradually increased blood pressure, it also favored an increasingly more elaborate clotting system. In other words, what is currently an irreducibly complex system could arise through a step-by-step evolutionary process along with the step-by-step coevolution of related systems.

Lastly, even if the claims of intelligent design advocates were plausible, the conclusion that an intelligent designer must have been responsible for irreducibly complex systems is not scientific. According to Eugenie Scott, the executive director of the National Center for Science Education:

> But even if natural selection were unable to explain the construction of irreducibly complex structures, does this mean that we must now infer that intelligence is required to produce such structures? Only if there are no other natural causes—known or unknown—that could produce such a structure. Given our current knowledge of the mechanisms of evolution, there is no reason why natural selection could not explain the assembly of an irreducibly complex structure—but it is also the case that a future researcher might come up with an additional mechanism or mechanisms that could explain irreducibly complex structures by some other natural process.

Some scientists have described Behe's approach as an "argument from ignorance" because the intelligent creator is used as an explanation when a natural explanation is lacking. This is reminiscent of the "God of the gaps" argument, where

God's direct action is called upon to explain something that science has not yet explained. "God of the gaps" arguments are rejected by both theologians and scientists.[30]

The first court case challenging intelligent design took place in Dover, Pennsylvania, in 2005. The Dover school board had required science teachers to read a statement offering intelligent design as a contradictory alternative to Darwinian evolution. The case received national attention with prominent expert witnesses, among them Michael Behe supporting intelligent design and Kenneth Miller supporting evolution. In the end, the judge sided with the majority of scientists, concluding that intelligent design is "nothing less than the progeny of creationism"[31] and that it "is not science and cannot be adjudged a valid, accepted scientific theory."[32]

In spite of this legal setback, the intelligent design movement is strong, vocal, politically popular, and financially powerful. It is centralized in the Center for Science and Culture (CSC) of the Discovery Institute, headquartered in Seattle.[33] The movement's success is due in large part to its healthy funding and the academic credentials of its fellows and advisers.[34]

HOW SHOULD EVOLUTION BE TAUGHT IN PUBLIC SCHOOLS?

In democratic societies, people are free to think, speak, and believe as they wish. Historically and currently, the major question regarding evolution is not what people should or should not believe, but how evolution should be taught in the science curricula of public schools. Evolution is a central theme in biology and, therefore, an integral part of modern science. As such, it merits a prominent place in science curricula. The Supreme Court has ruled that laws prohibiting the teaching of evolution, as well as those requiring the teaching of creation science alongside evolution in state-supported schools, are unconstitutional because they violate the establishment clause.

However, are there plausible *scientific* alternatives to evolution that should also be taught as science alongside evolution? The creationist, creation-science, and intelligent design movements are all attempts to displace evolution or contradict it with so-called alternative scientific theories. All have failed, both scientifically and legally, as science. The current evidence supporting evolution is so overwhelming that denying it is the intellectual equivalent of denying gravity. Indeed, there is substantial scientific discussion about how evolution works, just as there is substantial scientific discussion about how gravity works. In the science classroom, we must not devote time to ideas that are not scientific.

While recognizing that evolution is indisputable given the overwhelming evidence supporting it, we cannot use it in the science classroom to either deny or confirm the existence of a supreme creator. Although some scientists deny the existence of God, such a denial is a belief, not a scientific conclusion, just as accepting the existence of God is a belief and not a scientific conclusion. Discussions of faith, agnosticism, and atheism find their proper place in philosophy, history, sociology, politics, literature, the arts, religion, and books (such as this one), but not in science education.

NOTES

1. The title of this chapter is derived from the title of an article by Notre Dame University professor Alvin Platinga, "When Faith and Reason Clash: Evolution and the Bible," *Christian Scholar's Review* 21, no. 1 (1991): 8–33. Platinga derived his title from the couplet "When faith and reason clash, Let reason go to smash," from a poem by Scottish bard William E. McGonagall, who has the unfortunate distinction of being dubbed the "writer of the worst poetry in the English language,"

http://www.mcgonagall-online.org.uk (accessed December 23, 2006). Platinga's article has sparked an interesting dialogue regarding this couplet. After quoting McGonagall, Platinga rephrases the couplet to reflect changing attitudes resulting from advances in science in the nineteenth century to "When faith and reason clash, Tis faith must go to smash." In response to Platinga's article, Howard Van Till further rephrased the couplet, "When faith and reason appear to clash, Tis the appearance must go to smash," http://www.asa3.org/asa/dialogues/Faith -reason/CRS9-91VanTill.html (accessed December 23, 2006).

2. C. E. Darwin, *On the Origin of Species by Means of Natural Selection or the Preservation of Favoured Races in the Struggle for Life*, 6th ed. (London: John Murray, 1872), pp. 422–23.

3. Ibid., pp. 421–22.

4. Ibid., p. 428.

5. T. Dobzhansky, "Nothing in Biology Makes Sense Except in the Light of Evolution," *American Biology Teacher* 35 (1973): 125–29.

6. The account of the Scopes trial given in this section, including all quotations, is based on information at Douglas Linder's Scopes trial home page, hosted by the University of Missouri at Kansas City, and its associated links, http://www.law.umkc.edu/faculty/ projects/ftrials/scopes/scopes.htm (accessed May 30, 2005). This section also relies on information from E. C. Scott, *Evolution vs. Creationism: An Introduction* (Westport, CT: Greenwood Press, 2004), pp. 93–97.

7. J. C. Whitcomb and H. M. Morris, *The Genesis Flood: The Biblical Record and Its Scientific Implications* (Philadelphia: Presbyterian and Reformed Publishing Co., 1961).

8. For detailed information on the Institute for Creation Research, consult its Web site at http://www.icr.org.

9. Scott, *Evolution vs. Creationism*, p. 106.

10. *McLean v. Arkansas Board of Education* as reprinted in *Science* 215 (1982): 934–43.

11. Ibid.

12. K. W. Miller, *Finding Darwin's God: A Scientist's Search for Common Ground between God and Evolution* (New York: Cliff Street Books, 1999), chap. 3.

13. For excellent, nontechnical reviews on the scientific basis for the age of Earth and the universe, see http://www.talkorigins.org/origins/faqs-youngearth.html.

14. Whitcomb and Morris, *Genesis Flood*, p. 369.

15. Miller, *Finding Darwin's God*, p. 80.

16. From the definition of "creation science" in the Arkansas law.

17. Whitcomb and Morris, *Genesis Flood*, p. 66.

18. C. Wieland, "Darwin's Finches: Evidence Supporting Rapid Post-Flood Adaptation," *Creation* 14, no. 3 (1992): 22–23.

19. Miller, *Finding Darwin's God*, p. 80.

20. Data from Gallup polls are available online (with subscription) from http://www.gallup.com. The results of this particular survey are summarized at numerous Web sites.

21. W. Paley, *Natural Theology* (London: J. Faulder, 1809), pp. 1–2.

22. P. E. Johnson, *Darwin on Trial* (Washington, DC: Regnery Publishing, 1991).

23. According to a Gallup poll, 95 percent of scientists support evolutionary theory. This poll included scientists in all fields of science and engineering; presumably the proportion among biologists is even greater.

24. Examples include C. M. Smith and C. Sullivan, *The Top Ten Myths about Evolution* (Amherst, NY: Prometheus Books, 2007); M. S. Shermer, *Why Darwin Matters: The Case against Intelligent Design* (New York: Times Books, Henry Holt & Co., 2006); M. Young and T. Edis, *Why Intelligent Design Fails: A Scientific Critique of the New Creationism* (Rutgers, NJ: Rutgers University Press, 2004); Scott, *Evolution vs. Creationism*; B. C. Forrest and P. R. Gross, *Creationism's Trojan Horse: The Wedge of Intelligent Design* (Oxford, UK: Oxford University Press); M. Perakh, *Unintelligent Design* (Amherst, NY: Prometheus Books, 2003); Miller, *Finding Darwin's God*; L. R. Godfrey, ed., *Scientists Confront Creationism* (New York: W. W. Norton & Co., 1983).

25. M. J. Behe, *Darwin's Black Box: The Biochemical Challenge to Evolution* (New York: Simon & Schuster, 1996).

26. Darwin, *Origin of Species*, p. 146.

27. Behe, *Darwin's Black Box*, p. 39.

28. Ibid., p. 39.

29. J. C. Eikenboom et al., "Multiple Substitutions in the von Willebrand Factor Gene that Mimic the Pseudogene Sequence," *Proceedings of the National Academy of Sciences, USA* 91 (1994): 2221–24.

30. Scott, *Evolution vs. Creationism*, p. 119.

31. *Kitzmiller et al. v. Dover Area School District*, Case 4:04-cv-02688-JEJ, Document 342 (2005), p. 31.

32. Ibid., p. 89.

33. For detailed information on the Discovery Institute, consult its Web site at http://www.discovery.org.

34. For a list of CSC fellows and the credentials of most of them, see http://www.discovery.org/csc/fellows.php.

Chapter 10

ABANDONING THE DICHOTOMY

I t is nothing short of spectacular that scientists have found so much evidence of evolution in so many diverse places—in fossils entombed in sequential layers of sedimentary rocks, in molecules of DNA, in the mutated forms of thousands of gene products, in the anatomy and physiology of organisms, in the inseparable links between species and their environments. Why, then, is there such vehement opposition to the very idea that life has evolved? Phillip Johnson, a founder of the intelligent design movement, claims that there is an irreconcilable dichotomy between religion and evolution:

> The story of salvation by the cross makes no sense against a background of evolutionary naturalism. The evolutionary story is a story of humanity's climb from animal beginnings to rationality, not a story of a fall from perfection. It is a story about recognizing gods as illusions, not a story about recognizing God as the ultimate reality we are always trying to escape. It is a story about learning to rely entirely on human intelligence, not a story of the helplessness of that intelligence in the face of the inescapable fact of sin.

162 RELICS OF EDEN

There is no satisfactory way to bring two such fundamentally different stories together, although various bogus intellectual systems offer a superficial compromise to those who are willing to overlook a logical contradiction or two. A clear thinker simply has to go one way or another.[1]

Under such a presumption, the thinking goes something like this: "If the Bible is literally true and inerrant then evolutionary theory must be false, regardless of the abundant evidence that supports it." A powerful Christian fundamentalist movement has embraced the dichotomy. Its adherents have organized a strong campaign against evolution from church pulpits, on radio and television broadcasts, in books and pamphlets, over the Internet, and through state legislation and mandates by school boards.

Religious fundamentalists are not the only ones who promote the dichotomy. On the opposite extreme are several well-respected biologists who claim that modern science refutes the existence of God. Perhaps the best known is Richard Dawkins, an Oxford University zoologist, eloquent humanist, and prolific author who has enthusiastically used evolution and natural selection to reason that God does not exist. He is one of the most controversial and quotable evolutionary biologists and, according to the promotional material for a recent book, "the world's most prominent atheist."[2] Although his antireligious views receive the most attention, much of his writing is enlightening and positive. Here's one of my favorite passages:

After sleeping through a hundred million centuries we have finally opened our eyes on a sumptuous planet, sparkling with colour, bountiful with life. Within decades we must close our eyes again. Isn't it a noble, an enlightened way of spending our brief time in the sun, to work at understanding the universe and how we have come to wake up in it? This is how I answer when I am asked—as I am surprisingly often—

why I bother to get up in the mornings. To put it the other way round, isn't it sad to go to your grave without ever wondering why you were born? Who, with such a thought, would not spring from bed, eager to resume discovering the world and rejoicing to be a part of it?[3]

Those with strong religious convictions, especially those who are already inclined to reject evolution, however, take offense at much of what Dawkins has written. For example, here are three often quoted passages in which Dawkins attempts to negate the premise of God as Creator:

Natural selection, the blind, unconscious, automatic process which Darwin discovered, and which we now know is the explanation for the existence and apparently purposeful form of all life, has no purpose in mind. It has no mind and no mind's eye. It does not plan for the future. It has no vision, no foresight, no sight at all.[4]

The universe we observe has precisely the properties we should expect if there is, at bottom, no design, no purpose, no evil and no good, nothing but blind, pitiless, indifference.[5]

Nearly all peoples have developed their own creation myth, and the Genesis story is just the one that happened to have been adopted by one particular tribe of Middle Eastern herders. It has no more special status than the belief of a particular West African tribe that the world was created from the excrement of ants.[6]

Even though Johnson and Dawkins are polar opposites in their claims, both argue that there is an irreconcilable dichotomy between evolution and religion. But is the dichotomy real?

MODERN PRINCIPLES OF SCIENCE
NEITHER DENY NOR PROMOTE RELIGION.

Creationist authors often stereotype scientists as materialistic atheists bent on using evolution to belittle religious faith. Some scientists are indeed outspoken atheists, whereas others openly promote their strong religious convictions. Many fall between these two extremes or are simply indifferent to religion.

A number of well-respected scientists, evolutionary biologists among them, have been devoutly religious. Sir Ronald Fisher, a founder of statistics and population genetics, was a staunch Darwinist. Yet he was a faithful Anglican throughout his life and often wrote religious articles for church magazines. Theodosius Dobzhansky, one of the world's most renowned evolutionary geneticists, helped pioneer the modern synthesis of Mendelian genetics with Darwinian natural selection. He was also a lifelong devotee of the Russian Orthodox Church. His famous article titled "Nothing in Biology Makes Sense Except in the Light of Evolution" has been ridiculed by intelligent design advocates because of its title.[7] I often wonder if those who disparage the title have actually read the article, which is intended to *promote* religious faith. In it he wrote: "It is wrong to hold creation and evolution as mutually exclusive alternatives. I am a creationist *and* an evolutionist. Evolution is God's, or Nature's, method of creation. Creation is not an event that happened in 4004 BC; it is a process that began some 10 billion years ago and is still under way."[8]

Francis Collins, director of the Human Genome Project, was once an atheist but is now a devout Christian. In his book *The Language of God*, Collins wrote:

Many people who have considered all the scientific and spiritual evidence still see God's creative and guiding hand at work. For me, there is not a shred of disappointment or disillusionment in these discoveries about the nature of life—quite the contrary! How marvelous and intricate life turns out to be! How deeply satisfying is the digital elegance of DNA! How aesthetically appealing and artistically sublime are the components of living things, from the ribosome that translates RNA into protein, to the metamorphosis of the caterpillar into butterfly, to the fabulous plumage of the peacock attracting his mate! Evolution, as a mechanism, can be and must be true. But that says nothing about the nature of its author. For those who believe in God, there are reasons now to be more in awe, not less.[9]

Fisher, Dobzhansky, and Collins are just three of the scientists who, for almost a century and a half of Darwinism, have found no difficulty maintaining their religious faith while embracing the reality of evolution. However, nearly all scientists, whether or not religious, forcefully oppose the creationist and intelligent design movements, which they brand as false claims to science.

A number of major scientific organizations concur that evolution is central to modern biology. At the same time, they have taken steps to ensure that their policies neither support nor deny the involvement of a creator in the origin and evolution of life.[10]

Many people have a serious misconception. They assume that science denies God's involvement in anything explainable purely by natural law. For example, Phillip Johnson wrote, "Evolutionary science will allow God to 'exist,' but it will not allow him to leave evidence of his existence in the realm accessible to science."[11] In fact, the principles that guide modern science neither ascribe natural law to God nor deny the possibility that God may work through natural law. Instead, belief in God (or in an intelligent designer) is a personal belief that can be every bit as compelling as the principles of science. *But such beliefs are not science and should not be taught as such.*

DOES EVOLUTIONARY SCIENCE DESTROY FAITH?

Creationist attacks on evolution are, for the most part, a reaction to fears that learning about evolution destroys religious faith. Although some people may abandon faith while studying evolution, such fears are largely unjustified, at least when considering the United States population as a whole.[12] A 2001 Gallup poll showed that 45 percent of Americans agreed with the statement that "God created human beings pretty much in their present form at one time within the last 10,000 years or so," thus rejecting the scientific evidence that the human species is far more ancient. Thirty-seven percent agreed with the statement that "human beings have developed over millions of years from less advanced forms of life, but God guided this process." Only 12 percent agreed that "human beings have developed over millions of years from less advanced forms of life, but God had no part in this process," and 6 percent claimed another belief or no opinion.

Therefore, at least 82 percent of Americans believe that God created us. The Gallup Organization administered this poll five times since 1982. Each time the results were essentially the same despite the astonishing advances in evolutionary science in the same period.[13]

Alexis de Tocqueville predicted this high proportion of religious Americans in 1835 when he wrote in *Democracy in America*,

> Never will the short space of sixty years contain all of a person's imagination; the incomplete joys of this world will never be enough for the heart. Alone among all beings, people display a natural distaste for existence and an immense desire to exist: they despise life and fear oblivion. These different instincts incessantly urge their souls toward contemplation of another world, and it is religion that leads them there. Religion is, therefore, just one particular form of hope, and it is as natural to the human heart as hope itself. It is by a kind of aberration of intelligence, and through a sort of moral vio-

lence imposed upon their own nature, that people abandon their religious beliefs; an invincible inclination brings them back. Incredulity is an accident; faith alone is the permanent state of humanity.[14]

Research on religious beliefs confirms Tocqueville's assertion. The staggering accumulation of evidence that supports evolution has done little to destroy the religious faith of Americans in general.

Most mainstream Christian, Jewish, and Islamic denominations or groups do not reject the possibility of evolution as an explanation of natural history, including human origins, as long as God is not explicitly denied. For example, in 1996, Pope John Paul II reaffirmed the 1950 Encyclical *Humani generis* of Pope Pius XII, stating that there is "no opposition between evolution and the doctrine of the faith about man and his vocation, on condition that one did not lose sight of several indisputable points."[15] Clearly, the evidence of evolution is not necessarily a serious threat to religious faith.

WHAT IS THE REAL DANGER?

I am dismayed over how often the authors of antievolution books misrepresent science. I can understand how a minister or a parent with little scientific training could oppose evolution on religious grounds. But many authors of antievolution literature are well educated in the sciences, and the claims they make in their books are, for the most part, unsupported by scientific evidence. What could be their motivation? I suspect that most of them truly believe they are engaged in a noble cause. Once they accept the evolution-creation dichotomy as real, they seem willing to paint an extremely selective picture of science, even misrepresent it, in the hope that viewers of their work will question science instead of faith. In other words, the end justifies the means.

The irony here is that such an effort may do more to *harm* faith than to promote it. Especially vulnerable are college and university students. Several surveys show that a significant proportion of students enter their college years accepting the dichotomy. Although not well informed about evolution, they already reject it.[16] A general biology course is a standard requirement at colleges and universities, and professors who teach such courses typically present abundant evidence of evolution along with the analytical skills students need to understand the evidence. Any preconceived notions that the scientific approach is weak or wrongheaded get shattered. Students quickly acquire information and discard the unsupported claims of creationists and intelligent design advocates. Recalling the propaganda about a dichotomy, they may end up questioning their faith.

The controversy stems from the nature and traditions of science and religion. Both are ways of searching for answers to help us make sense of the world in general, and human life in particular. However, the types of answers people hope to find from each of them differ. All major religions are spiritual; they transcend the material world. The founding texts of the major Western religions—the Bible, the Torah, and the Quran—all state that God created the earth. Their accounts of creation are mythical in the most positive, wondrous sense of the word. They contain a wealth of information on the purpose of creation but virtually nothing on the mechanisms that brought it about. By contrast, science deals entirely with the mechanisms. It is and must remain silent on the spiritual purpose of life.

Modern science is absolutely incompatible with the creationist views of a universe less than ten thousand years old, a worldwide flood that obliterated all terrestrial life except that preserved on an ark, and the special creation of all species including humans. For those who still insist on such a narrow interpretation of religious texts, there is indeed a dichotomy.

Is such a strict literal view really necessary? The debate over how literally religious texts should be interpreted is a very old one, and some of the best advice comes from several of the earliest Christian theologians. For example, Origen of Alexandria lived from about 185 to 254 CE, and many consider him to be the first Christian philosopher. Referring to the biblical creation account in Genesis, Origen wrote this:

> Now what intelligent person would believe that the first, second, and third day took place without a sun, moon, or stars—indeed, the first day, as it were, without a heaven? Who could be so childish as to think that God was like a human gardener and planted a paradise in Eden facing the east, and in it made a real visible tree, so that one could acquire life by eating its fruit with real teeth, or, again, could participate in good and evil by eating what he took from the other tree?

> And if the text says that God walked in the garden in the evening, or Adam hid himself under the tree, I cannot think that anyone would dispute that these things are said in a figurative sense, in an effort to reveal certain mysteries by means of an apparent historical tale and not by something that actually took place.[17]

About a century and a half later, Augustine (354–430 CE) wrote *The Literal Meaning of Genesis*. Much of this work is devoted to the question of how literally we should interpret the first three chapters of Genesis. In the opening paragraph, he wrote,

> In all the sacred books, we should consider the eternal truths that are taught, the facts that are narrated, the future events that are predicted, and the precepts or counsels that are given. In the case of a narrative of events, the question arises as to whether everything must be taken according to the figurative sense only,

or whether it must be expounded and defended also as a faithful record of what happened. No Christian will dare say that the narrative must not be taken in a figurative sense.[18]

Sixteen centuries after Augustine penned these words, millions of people willingly reject scientific evidence on the basis of a strictly literal interpretation of scripture. If instead, however, we view religious texts as spiritual guiding documents rather than literal historical records, the perceived dichotomy fades. Religion and science become complementary ways to interpret our world.

Those who sincerely seek both scientific and spiritual understanding would do well to abandon the dichotomy. Denying the evidence of evolution, including human evolution, is honest only in ignorance. The incredible diversity of life on Earth, the many fossils unearthed, the varied yet similar anatomical features among species, the obvious hierarchical arrangement of life, and the literally millions of ancestral relics in our DNA—all undeniably attest to our common evolutionary origin with the rest of life. If someone can believe that all living organisms share the same creator, why not consider that all living organisms share a common genetic heritage? Indeed, we can find wonder, even comfort, in embracing our biological relationship with all living things. As Darwin understood, "there is grandeur in this view of life."

NOTES

1. P. E. Johnson, *Defeating Darwinism by Opening Minds* (Downer's Grove, IL: InterVarsity Press, 1997), p. 111.

2. See http://richarddawkins.net/godDelusion (accessed December 28, 2006).

3. R. Dawkins, *Unweaving the Rainbow: Science, Delusion and the Appetite for Wonder* (Boston: Houghton Mifflin, 1998), p. 6.

4. R. Dawkins, *The Blind Watchmaker: Why the Evidence of Evolution Reveals a Universe without Design* (New York: W. W. Norton & Co., 1986), p. 5.

5. R. Dawkins, *River Out of Eden* (New York: Basic Books, 1995), p. 133.

6. Dawkins, *The Blind Watchmaker*, p. 316.

7. For example, see J. Wells, *Icons of Evolution: Why Much of What We Teach about Evolution Is Wrong* (Washington, DC: Regnery Publishing, 2002), pp. 245–48.

8. T. Dobzhansky, "Nothing in Biology Makes Sense Except in the Light of Evolution," *American Biology Teacher* 35 (1973): 125–29.

9. F. S. Collins, *The Language of God: A Scientist Presents Evidence for Belief* (New York: Free Press, 2006), pp. 106–107.

10. See National Association of Biology Teachers Web site, http://nabt.org/sub/position_statements/evolution.asp (accessed November 24, 2005).

11. Johnson, *Defeating Darwinism*, p. 121.

12. Scientists, including those in both natural sciences and social sciences, are less likely to be religious when compared to the general public, although it is not certain that this is due specifically to the study of evolution.

13. Data from Gallup polls are available online (with subscription) from http://www.gallup.com. The results of this particular survey are summarized on numerous Web sites.

14. A. de Tocqueville, *De la Démocratie en Amérique, I, Deuxième Partie,* http://classiques.uqac.ca/classiques/De_tocqueville_alexis/democratie_1/democratie_t1_2.pdf, p. 122 (accessed December 25, 2006). The English translation is my own from the original French: "Jamais le court espace de soixante années ne renfermera toute l'imagination de l'homme; les joies incomplètes de ce monde ne suffiront jamais a son cœur. Seul entre tous les êtres, l'homme montre un dégoût naturel pour l'existence et un désir immense d'exister: il méprise la vie et craint le néant. Ces différents instincts poussent sans cesse son âme vers la contemplation d'un autre monde, et c'est la religion qui l'y conduit. La religion n'est donc qu'une forme particulière de l'espérance, et elle est aussi naturelle au cœur humain que

l'espérance elle-même. C'est par une espèce d'aberration de l'intelligence, et à l'aide d'une sorte de violence morale exercée sur leur propre nature, que les hommes s'éloignent des croyances religieuses; une pente invincible les y ramène. L'incrédulité est un accident; la foi seule est l'état permanent de l'humanité."

15. Pope John Paul II, "Message to the Pontifical Academy of Sciences," *Quarterly Review of Biology* 72, no. 4 (1997): 382.

16. For a review of surveys, see E. C. Scott, "Problem Concepts in Evolution: Cause, Purpose, Design, and Chance," National Center for Science Education, October 1, 1999, http://www.ncseweb.org/resources/articles/695_problem_concepts_in_evolution_10_1_1999.asp (accessed December 30, 2006).

17. H. A. Musurillo, *The Fathers of the Primitive Church* (New York: New American Library, Inc., 1966), p. 202.

18. Augustine, *The Literal Meaning of Genesis*, trans. J. H. Taylor, SJ, in *Ancient Christian Writers: The Works of the Fathers in Translation*, no. 41, ed. J. Quasten, W. J. Burghardt, and T. C. Lawler (New York: Newman Press, 1982), p. 19. I highly recommend this volume of Augustine's writings for those interested in how one of Christianity's earliest and most influential scholars interpreted the first three chapters of Genesis in light of what was known in his day. Given the title of this volume, some might find it quite surprising how Augustine repeatedly argues *against* literal interpretations, frequently offering a series of difficult questions and alternative interpretations of biblical passages.

INTRODUCTION TO THE APPENDICES

A n old proverb, attributed to Gustave Flaubert (1821–80), states, *"Le bon Dieu est dans le détail."* It has been variously restated and revised as "God is in the details," "The devil is in the details," and "The truth, if it exists, is in the details." Regardless of how it is phrased, the proverb's message is clear: details are crucial. To me, the most stunning elegance in the evidence of evolution resides in the technical details. They are much like the tool marks in Michelangelo's unfinished marble statues, the subtle interplay of warm and cool hues in a painting by Vermeer, or the intricate harmonies in the orchestral accompaniment of a Puccini aria, often overlooked but absolutely essential. While writing this book, I often had to resist the temptation to include technical details that magnificently confirm the evidence presented in each chapter but that could easily overwhelm readers who are not well versed in molecular biology. For those of you who relish the details, however, the first two appendices should at least whet your appetites, if not satisfy them.

These two appendices highlight especially intriguing examples covered in the main text of the book but in much greater detail. They are loosely, but not entirely, tied to specific chapters in the book, often integrating topics from several chapters. They are best read after one has finished the main text of the book.

Appendix 1 tells the story of the *NANOG* gene and its pseudogenes. I chose this example for three reasons. First, it nicely illustrates a gene, a duplication pseudogene, ten retro-pseudogenes, multiple retroelements, mutations, and natural selection—all major topics in the book—in a single, integrated package, and shows how they are related to one another. Second, as an essential gene for maintenance of human embryonic stem cells, the *NANOG* gene is currently one of the hottest genes in scientific research. Third, part of my own research is focused on the *NANOG* gene and its pseudogenes, so I am well acquainted with many of the finest details.

Appendix 2 highlights the nine inversions that distinguish human and chimpanzee chromosomes. Chapter 1 explored the details of the chromosome 2 fusion that distinguishes these two species; appendix 2 follows up with the details of the inversions. Together, chapter 1 and appendix 2 cover the details of all major chromosomal rearrangements that differ between the human and chimpanzee genomes.

I considered writing an appendix on the details of molecular evidence from mitochondrial and Y-chromosome DNA supporting the ancient human dispersals depicted in figure 7.1. However, Spencer Wells, director of the Genographic Project, has already described the DNA evidence in greater detail than I would have done—and far more authoritatively and eloquently—in the appendix of his book *Deep Ancestry*.[1] For readers interested in DNA evidence of human migrations, I highly recommend this book and the Genographic Project's "Atlas of the Human Journey" Web site at http://www.nationalgeographic.com/genographic/atlas.html.

Nearly all of the evidence presented in *Relics of Eden* is recent, much of it discovered within the past few years. However, these discoveries rest entirely on a vast foundation established by brilliant and dedicated scientists who, for almost a century and a half since the publication of the *Origin of Species*, pioneered the science of evolutionary genetics. The scientific journey from Darwin's *Origin of Species* to the genome era of today contains some of the most compelling stories in all of history. Appendix 3 departs from the detailed nature of the first two appendices and offers a very brief history of the major landmarks in evolutionary genetics. The focus is on scientists and their major contributions, with almost no detail of their experiments and reasoning. For those who want more detail on these scientists, I recommend reading their biographies in books, which can be found in most libraries, or brief biographies on the Internet.

The chapters and appendices of this book are mere introductions to the vast information available on the evidence of evolution in human DNA. For more information, I recommend exploring sources in the bibliography that follows the appendices.

NOTE

1. S. Wells, *Deep Ancestry: Inside the Genographic Project* (Washington, DC: National Geographic Society, 2006).

Appendix I

THE STORY OF NANOG AND ITS PSEUDOGENES

Chapters 5 and 6 highlight the *NANOG* gene and its pseudo-genes because they so beautifully exemplify the evolutionary relics that this book is about—genes, retroelements, duplication pseudogenes, retropseudogenes, and the effect of natural selection—all in one package. We are about to see how these different lines of evidence within the same story are fully consistent with one another.

HOW DO WE DISTINGUISH DUPLICATION PSEUDOGENES FROM RETROPSEUDOGENES?

Of the eleven *NANOG* pseudogenes in humans, ten are retro-pseudogenes and one is a duplication pseudogene. How can we tell the difference between these two classes of pseudogenes? One of the key components for transposition of retropseudo-genes (as well as retroelements) is an enzyme called reverse transcriptase. You may recall that transcription is the process of copying the sequence in DNA into RNA, and transcription is essential for our genes to be expressed. Typically, enzymes (called RNA polymerases) transcribe a gene to make an RNA copy of it, and then the RNA copy is translated into a protein.

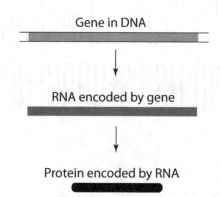

As its name implies, reverse transcriptase does just the reverse of transcription—it uses an RNA molecule as a template to make a DNA copy. The prefix *retro-*, as in "retrovirus," "retropseudogene," and "retroelement," is intended to imply the ability of this enzyme to retrocopy RNA into DNA.

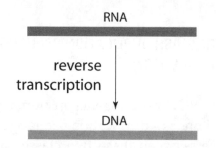

Retroviruses, like influenza viruses and HIV, carry their own reverse transcriptase gene. When they enter a cell through infection, their reverse transcriptase gene can be translated directly from the virus's RNA into the reverse transcriptase enzyme. The enzyme then makes a DNA copy of the virus's

RNA to insert into the cell's DNA. The virus DNA, now integrated into the cell's DNA, can be transcribed into multiple RNA copies, which then assemble themselves into intact viruses and go on to infect other cells.

Certain retroelements present in our genome also contain their own reverse transcriptase genes. Many of these retroelements are incomplete because reverse transcriptase often stops short of making a complete DNA copy from the RNA. These incomplete retroelements can still be inserted into the DNA and most of the retroelements in our genome are incomplete pieces of the full-length versions. Even so, we still have thousands of copies of complete retroelements with intact reverse transcriptase genes.

The reverse transcriptase enzymes encoded by these genes can retrocopy any RNA molecule into DNA. Most of the RNAs that are retrocopied into DNA are the retroelements themselves. However, because all active genes encode RNA, the RNA of genes can be retrocopied into DNA and inserted into the genome as retropseudogenes. This is how our genome acquired so many retropseudogenes, which are simply DNA copies of reverse-transcribed RNAs.

Because reverse transcriptase copies the RNA from genes to make pseudogenes, we can distinguish retropseudogenes from other types of pseudogenes by their lack of introns. Very soon after an RNA molecule is transcribed from a gene, the introns are removed from the RNA. Since most retropseudogenes arise from RNAs that no longer have their introns, retropseudogenes also lack the introns.

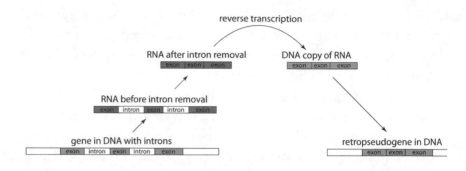

Furthermore, because retropseudogenes are derived from RNA, all genes in the genome should not be represented equally among the retropseudogenes. Retropseudogenes can be produced in almost any cell. However, for a newly formed retropseudogene to be passed on from parent to child, it must originate in a *germline cell*—a cell that will eventually give rise to egg or sperm cells. Any retropseudogene that arises in one of the trillions of cells in a human body that are not part of the germline does not survive beyond the life of the person who carries it.

For this reason, only those genes expressed in the germline produce inherited retropseudogenes, and we often have multiple copies of them in our genomes. The *NANOG* gene is expressed in embryonic stem cells, which are produced in the first few days of an embryo's life. Every cell in your body traces its ancestry to a tiny mass of embryonic stem cells, each of which had its *NANOG* gene activated at that time. These embryonic stem cells differentiated into all tissues of the body, including the germline, so it is no surprise that there are multiple retropseudogene copies of the *NANOG* gene in our genome.

By contrast, hemoglobin is one of the most abundant proteins in our body—it is present in every red blood cell throughout all of the blood. Thus, the genes that encode hemoglobin produce an enormous amount of RNA. Offhand, it might seem that we should have many retropseudogene copies of the hemoglobin-encoding genes because of the abundant RNA available to be retrocopied. In fact, as we saw in chapters 2 and 3, there are several pseudogene copies of the genes that encode hemoglobin, but all of them are duplication pseudogenes, with the introns still present. Not a single retropseudogene of these hemoglobin-encoding genes is in our genome. This observation makes sense when we realize that the genes encoding hemoglobin are expressed in the bone marrow. Germline cells contain these genes, but the genes are shut off; they produce no RNAs, so they are not retrocopied into retropseudogenes in the germline.

The lack of introns in retropseudogenes is strong evidence that they arose from reverse transcription of RNA into DNA. However, lack of introns is not the only evidence of reverse transcription. RNAs transcribed from protein-encoding genes have yet another unique feature that tells us how they originated. After an RNA is transcribed from a gene, an enzyme adds a string of As to the end of it. The string of As is called a *poly-(A) tail*. There is no string of A-T pairs in the DNA of a gene to encode the poly-(A) tail in the RNA; instead the enzyme simply adds the As after the RNA has dissociated itself from the gene. If retropseudogenes are reverse transcribed from RNA, they, too, should have a poly-(A) tail on their ends. Indeed a poly-(A) tail is evident in many of them. I have to use the word *many* because some retropseudogenes are not full-length copies of the gene, just a part of it, and sometimes the missing part is on the end that has the poly-(A) tail. Duplication pseudogenes, on the other hand, should not have a poly-(A) tail because they are copied from the DNA itself, and genes in DNA have no poly-(A) tails. Indeed, duplication pseudogenes lack a poly-(A) tail.

Of the human *NANOG* pseudogenes, *NANOGP2*, *P4*, *P5*, *P7*, *P8*, *P9*, *P10*, and *P11* have poly-(A) tails and they lack introns. Thus, they are retropseudogenes. *NANOGP1* has introns and it has no poly-(A) tail, so it is a duplication pseudogene. It is also located very close to the *NANOG* gene, which is typical of most duplication pseudogenes because they often arise through tandem duplication of the segment of DNA that contains them. *NANOGP3* and *P6* lack poly-(A) tails because they are not full-length pseudogenes and they are missing the end onto which the poly-(A) tail is attached. They can still be identified as retropseudogenes because they lack introns at the sites where the introns are present in the *NANOG* gene.

Ancient retropseudogenes, not surprisingly, often have a mutated poly-(A) tail with Gs, Cs, and Ts interspersed among the As. The poly-(A) tail, along with the rest of a retropseudogene, is completely nonfunctional, so any mutations in it are selectively neutral. The older the pseudogene is, the more mutations we expect to see in its poly-(A) tail. Because the *NANOGP8* pseudogene is relatively young, its poly-(A) tail is still unmutated with a string of twenty-six As:

...AAAAAAAAAAAAAAAAAAAAAAAAAA...

The *NANOGP5* pseudogene, on the other hand, is quite old and its poly-(A) tail, although mutated, is still easily identifiable by its position on the end of the pseudogene and the relatively high number of As in it:

...AAAAAAAAAAAAAAGCAGCCAAAGAAAAA...

RETROTRANSPOSITION GENERATES REPEATS ON BOTH SIDES OF A RETROELEMENT OR RETROPSEUDOGENE.

The cellular mechanism that inserts retroelements and retropseudogenes into DNA duplicates a portion of the DNA at the insertion site so that retroelements and retropseudogenes are usually flanked by short repeats of DNA. As shown in figure A1.1, when a retroelement or retropseudogene is inserted into

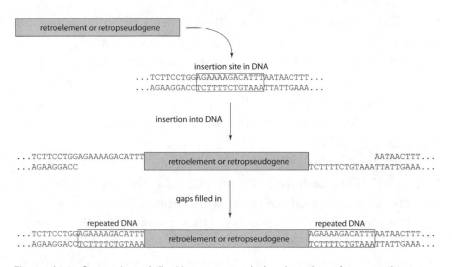

Figure A1.1. Generation of flanking repeats during insertion of a retroelement or retropseudogene.

the DNA, the DNA is cut in a staggered fashion, leaving short segments of single-stranded DNA on both sides of the inserted DNA. Enzymes then fill in the gaps to make the DNA fully double-stranded.

Thus, when we find recently inserted retroelements or retropseudogenes in DNA, they should have repeated segments on both sides as relics of the insertion event. The word *recently* is required here because the repeats on both sides of very ancient retroelements and pseudogenes are often mutated so that they are no longer identical to each other. Incidentally, *recently* in this case means within about the past ten million years—remember that we're speaking about an evolutionary timescale of perhaps a hundred million years or more.

In chapter 5, we saw that the *NANOG* gene has eight *Alu* elements (either full-length or truncated) in its introns. It also has a ninth *Alu* element that happens to be in one of the exons, and it's the one we discussed in chapter 6. This *Alu* element is in a region of the gene called the *3´ untranslated region*, or *3´*

UTR. Remember that genes in DNA consist of both exons and introns. The introns are removed after the RNA is made, leaving only the exons spliced to one another. The part that encodes the protein is somewhere in the middle of the RNA and is called the reading frame. A segment called the *5´ untranslated region*, or *5´ UTR*, precedes the reading frame, and following the reading frame is the 3´ UTR.

Mutations, including transposable elements, in either the 5´ UTR or 3´ UTR, usually have no effect on the protein because these regions lie outside of the reading frame. Therefore, a transposable element in the 3´ UTR, like the one in the human *NANOG* gene, is selectively neutral and natural selection does not remove it from the gene. Figure A1.2 depicts the human *NANOG* gene and the RNA it encodes, showing the positions of its nine *Alu* elements, drawn to scale. Notice that the single *Alu* element in the 3´ UTR remains in the RNA, whereas the other *Alu* elements are removed along with the introns that carry them.

As we saw in chapter 6, the chimpanzee *NANOG* gene has this same *Alu* element in exactly the same position in the 3´ UTR as in the human gene. Therefore, the *Alu* element must have been inserted into the *NANOG* gene before the lineages leading to humans and chimpanzees diverged. What do we see if we compare the human and chimpanzee *NANOG* genes to the rhesus macaque *NANOG* gene? I compared the DNA of the *NANOG* gene from all three species and discovered that the *Alu*

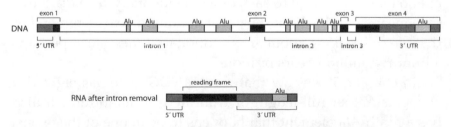

Figure A1.2. Structures of the human and chimpanzee *NANOG* gene in DNA and the RNA encoded by the gene after intron removal. An *Alu* element remains in the 3´ UTR in the RNA.

element present in the 3′ UTR of the human and chimpanzee versions of the gene is missing from the rhesus macaque version, suggesting that the *Alu* element was inserted *after* the lineage leading to rhesus macaque diverged from the lineage leading to humans and chimpanzees but *before* the human-chimpanzee divergence.

If this suggestion is correct, we should be able to find the site in the rhesus macaque *NANOG* gene that corresponds to the site where the *Alu* element was inserted into the human and chimpanzee versions of the *NANOG* gene. Also, part of that DNA should be repeated on both sides of the *Alu* element in the human and chimpanzee versions as relics of the insertion event. As shown in figure A1.3, this is exactly what we see when we compare the DNA sequences in these three species.

Now, let's test for consistency of the evolutionary history among these different lines of evidence. According to our discussion in chapter 6, the *NANOGP8* pseudogene arose exclusively in the lineage leading to humans after that lineage diverged from the one leading to chimpanzees. *NANOGP8* is a retropseudogene, so it should lack the introns that are present in the *NANOG* gene, and indeed it does. However, the 3′ UTR remains in the RNA encoded by the gene, so the *NANOGP8*

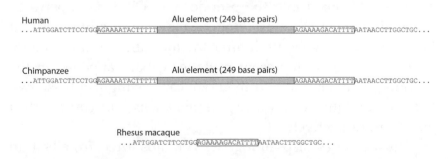

Figure A1.3. Portion of the 3′ UTR of the *NANOG* gene in humans, chimpanzees, and rhesus macaques. Repeats flank the *Alu* element in the 3′ UTR of the human and chimpanzee versions of the *NANOG* gene. The same site in the rhesus macaque version of the *NANOG* gene lacks the *Alu* element and has only one copy of the sequence that is repeated in humans and chimpanzees. The left and right repeats differ slightly due to mutations.

pseudogene should contain the *Alu* element. Indeed, the *Alu* element is right where it should be in the *NANOGP8* pseudogene.

TWO METHODS REVEAL THE RELATIVE AGES OF THE *NANOG* PSEUDOGENES AND A SURPRISING FEATURE OF THE *NANOGP1* DUPLICATION PSEUDOGENE.

A fairly simple and straightforward way to determine the relative ages of pseudogenes derived from the same gene is to compare all of them with the functioning gene. Because mutations in pseudogenes are selectively neutral, they should accumulate essentially at random without being eliminated by natural selection. Thus, the more mutations a pseudogene has when compared to the functioning gene, the older it must be.

This logic is based on an assumption that is not entirely correct, however. We have to recognize that the sequence in the modern gene may have changed over evolutionary time. When we compare functional genes in different species, the genes tend to be similar but not identical, evidence that mutations happened in them after the divergence of the lineages leading to these species. When a gene generates a pseudogene, its sequence is copied into the pseudogene. Differences between the modern gene and a pseudogene may reflect mutations that happened in the gene itself after the formation of the pseudogene, or they may represent mutations that happened in the pseudogene after it was formed. We can distinguish these two types of mutations as *source-gene mutations* and *post-insertion mutations*, respectively.

Thus, a series of pseudogenes contains a record, albeit an imperfect one, of how the gene itself has evolved over time, because the pseudogenes contain source-gene mutations. Another method of determining the relative ages of pseudogenes is to identify the source-gene mutations and then find the order of pseudogenes that requires the smallest number of source-gene mutations to explain the order. Such a method is based on the assumption of *parsimony*—the order requiring the fewest number of mutations is most likely to represent the actual history of pseudogene insertion.

We applied this method to the *NANOG* gene and its pseudogenes in the human and chimpanzee genomes and, not surprisingly, came up with almost the same order as the simpler method of comparing the pseudogenes with the modern gene.[1] There was one discrepancy. The simpler comparison method places *NANOGP1* as being younger than *NANOGP4*, whereas the source-gene mutation method places these two pseudogenes in the reverse order.

Fortunately, data available in DNA-sequence databases allowed us to fully resolve this discrepancy. If pseudogenes are completely inactive, then mutations anywhere in them should be selectively neutral. Thus, we expect mutations to be distributed at random in pseudogenes, and in nearly all cases they are. However, we noticed that nearly all of the mutations in *NANOGP1* were concentrated near the beginning of the gene, clearly a nonrandom distribution. What could explain this curious pattern?

Scientists can isolate and purify RNA molecules transcribed from genes and determine their sequences. Many copies of RNA transcribed from the human *NANOG* gene have been isolated from embryonic stem cells, then purified and sequenced, and their sequences deposited in DNA databases available via the Internet. Interestingly, among the many *NANOG* RNA sequences in the database, two were derived not from the functional gene but from *NANOGP1*, a duplication pseudogene, and they revealed a curious feature. The introns had been removed from the RNA, but the first intron is not the same in *NANOG* and *NANOGP1*. The segment that contains most of the mutations in *NANOGP1*, it turns out, lies within a segment removed as an intron from the RNAs produced by *NANOGP1* but present as an exon in the RNAs produced by *NANOG*. In other words, most of the mutations that distinguish *NANOGP1* from *NANOG* are in an intron *unique* to *NANOGP1*.

This observation suggests a clear explanation: *NANOGP1* was not always a pseudogene. The fact that RNAs have been recovered from it suggests that even now it may still be active as a gene, although probably much less active than the *NANOG* gene as evidenced by the relatively few RNAs it produces. As an active gene, mutations in the protein-encoding region should be selectively relevant, and natural selection should remove most of them. However, mutations in an intron are selectively neutral and are not removed. The fact that most of the mutations that distinguish the *NANOG* from *NANOGP1* lie within an intron unique to *NANOGP1* suggests that natural selection has preserved the protein-encoding part of the *NANOGP1*. Thus, *NANOGP1* has fewer mutations in it than it would have accumulated had it been a pseudogene right from the start, and therefore it is actually older than it appears to be based on a simple comparison with the *NANOG* gene. The order revealed by source-gene mutation (*NANOGP1* older than *NANOGP4*) is, in all likelihood, the correct one.

NOTE

1. D. J. Fairbanks and P. J. Maughan, "Evolution of the *NANOG* Pseudogene Family in the Human and Chimpanzee Genomes," *BMC Evolutionary Biology* 6 (2006): 12, http://www.biomedcentral.com/1471-2148/6/12; H. A. F. Booth and P. W. H. Holland, "Eleven Daughters of *NANOG*," *Genomics* 84 (2004): 229–38.

Appendix 2

NINE INVERSIONS

A t the genetic level, the most compelling differences between closely related species, and even distantly related species, are chromosome rearrangements, including fusions, fissions, inversions, translocations, and major duplications or deletions. They, more than any other type of hereditary variation, are responsible for the genetic reproductive barriers that ultimately separate varieties of the same species into distinct species causing these species to continue their divergence over the course of evolutionary time. Therefore, they also provide some of the most informative evidence of evolutionary histories.

In recent years, two methods have allowed scientists to make astounding progress with cross-species chromosome comparisons. The first is a method called *fluorescence in situ hybridization*, more commonly known by the acronym *FISH*. In this method, scientists take DNA from a chromosome segment in one species, label the DNA with a fluorescent dye, separate the DNA into single strands, then add the labeled DNA to the intact chromosomes of another species in which the DNA in the chromosomes has been forced to separate into single strands.

The labeled DNA, which is called a *probe*, forms base pairs wherever it encounters highly similar DNA in the chromosome, a process called *DNA hybridization*. The scientists then view the chromosomes illuminated with ultraviolet light under a microscope. The dye fluoresces, revealing the position of the labeled probe in the chromosomes and, as a consequence, the position of the matching DNA in the chromosome. Probes that either span, or are located on either side of, a breakpoint for a chromosome rearrangement that differs in two species can reveal the differences in that rearrangement when the two species are compared.

The second method is straightforward comparison of DNA sequences between species. The most informative DNA sequences are typically very large. Here is where genome projects have contributed enormously to studies of evolutionary biology. The vast DNA-sequence databases containing genome-sequence information allow scientists anywhere to conduct research on evolutionary biology with a computer. Through DNA sequence analysis, scientists can pinpoint chromosome rearrangements directly within DNA sequences.

The most powerful approach, however, is a combination of these two methods. All of the studies highlighted in this appendix do just that—they combine both of these methods to reveal the details of the breakpoints of the nine inversions that distinguish human and chimpanzee chromosomes. In so doing, they establish not only the positions of these breakpoints but they reveal information on the evolutionary history of these inversions.

Simple comparison of inversions in two species does not reveal which species carries the inversion and which carries the original ancestral conformation. However, when scientists compare chromosomes from humans and chimpanzees with those from gorillas, orangutans, and Old and New World monkeys, they can determine, usually without question, which species carries the original ancestral conformation and which carries the inversion. For chromosomes 1 and 18, the inversion happened exclusively in the lineage leading to humans. The remaining seven inversions in chromosomes 4, 5, 9, 12, 15, 16, and 17 are specific to the chimpanzee ancestral lineage. Figure A2.1 shows the positions of these inversions in the human and chimpanzee chromosomes. Arrows pointing down represent the ancestral conformation and those pointing up the inversion. The bonobo has the same inversions as the chimpanzee, indicating that all seven chimpanzee-specific inversions arose

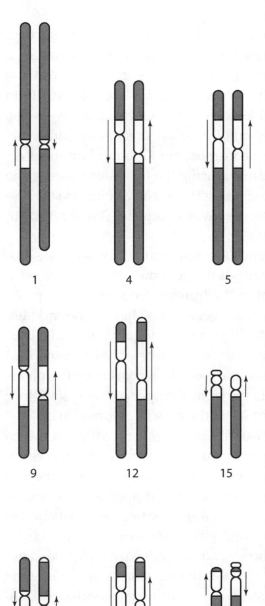

Figure A2.1. Positions of the nine inversions that distinguish human and chimpanzee chromosomes. Human chromosomes are on the left, chimpanzee on the right. Arrows pointing downward denote the ancestral conformation, arrows pointing up, the derived inversion.

before the divergence between the common chimpanzee and bonobo lineages.

How can a species live with an inversion? Shouldn't such an arrangement disrupt genes when the breakpoint is within a gene? The breakpoints of eight of the nine inversions did not disrupt any genes. For most inversions, the breakpoints lie in the vast amount of DNA between genes. Of the nine inversions we are about to examine, one inversion breakpoint did interrupt a gene in chimpanzee chromosome 12, but it had no effect on the gene's function because during the inversion process, the gene was duplicated so that a functional copy of it was still present after the inversion process was complete. (We'll return to this later in this appendix.)

Breakpoints within genes are uncommon for two reasons. First, only a small percentage of the genome consists of protein-encoding genes (2 percent in the human genome), so the probability of a breakpoint within a gene is low but not remarkably low. Second, a breakpoint within a gene disrupts the gene, usually inactivating the gene. If the gene is essential or beneficial (as most genes are), natural selection should disfavor the inversion, eventually eliminating it from the species, unless another copy of the gene is present (as is the case in chromosome 12).

How do inversions arise? There are several different mechanisms, and we will encounter three of them, with some variations, as we examine the nine inversions. However, one mechanism needs our attention now because it applies to so many inversions. Whenever there are large repeated segments in different DNA molecules, or even within the same DNA molecule, they may undergo a process called *recombination* or *crossing-over*, in which the DNA molecules line up side by side along the repeat and then exchange places within the repeated region. This process of crossing-over is a normal, common event between matching chromosomes, but on rare occasions it can happen between repeated segments on the same chromosome.

If the two repeats are in the same chromosome and are inverted relative to each other, recombination between them results in an inversion. Figure A2.2 explains the general process.

In this appendix, we'll look at each of the nine inversions in detail to see how they arose and what comparison of the human and chimpanzee DNA at the breakpoints tells us about the evolutionary history of the inversions. Most of the research sum-

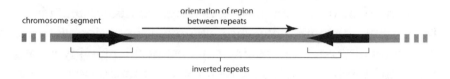

(a) A segment in a chromosome contains inverted repeats.

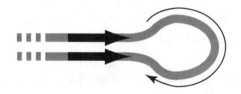

(b) Inverted repeats pair with each other, causing the chromosome to loop.

(c) The repeats recombine, inverting the region between them.

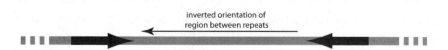

(d) An inversion is established in the chromosome.

Figure A2.2. Model explaining how inversions may arise by recombination between inverted repeats.

marized in this appendix was conducted at the University of Ulm in Germany under the direction of Dr. Hildegard Kehrer-Sawatzki.[1]

THE INVERSION IN CHROMOSOME 1
IS ASSOCIATED WITH HUMAN-SPECIFIC DUPLICATIONS.

Of the nine major inversions that distinguish human and chimpanzee chromosomes, the chromosome 1 inversion was the most difficult to define. It is the smallest of the nine inversions and both of its breakpoints are near the centromere, one on each side. Centromeres are often surrounded by repeated sequences, which is the case in chromosome 1, making localization of the precise breakpoints impossible with current methods. However, with extensive analysis, Szamalek et al. were able to define the breakpoints quite closely and determine how the inversion most probably occurred.[2] They published their analysis in 2006, making the chromosome 1 inversion the last of the nine to be documented in detail.

Looking at the comparison of human and chimpanzee chromosomes in figure A2.1, notice that the region where the inversion took place in chromosome 1 is larger in the human chromosome. The analysis of Szamalek et al. explains this discrepancy in size. They found that a segment still present in its original location in chimpanzee chromosome 1 was duplicated in the human lineage after it had diverged from the chimpanzee lineage and transposed to a position near the centromere in the human ancestral lineage. A portion of that sequence amounting to about ninety-one thousand base pairs was then duplicated again and transposed to the other side of the centromere but in the opposite orientation, leaving inverted repeats on either side of the centromere. Because these particular repeats were inverted relative to each other, recombination between them resulted in an inversion via the mechanism illustrated in figure A2.2. Chimpanzee chromosome 1 lacks these repeated segments and thus the corresponding region is smaller and also not likely to undergo an inversion.

The repeated region surrounding the centromere in human chromosome 1 is present in every human, and it confers an unfortunate disadvantage to us. Because of these large repeats, the region is prone to expansion with additional repeats, which can cause congenital abnormalities and mental retardation when the expansions happen. The repeated segment is also prone to breakage, which can contribute to cancer. Numerous cases of cancer are associated with chromosome breakage at this site.

Interestingly, this is not the first inversion in the evolutionary history of chromosome 1. When the human and great-ape versions of this chromosome are compared with their counterparts in monkeys, it is clear that a larger inversion happened in the human–great ape lineage after it diverged from the one that led to monkeys. All humans and great apes carry this larger inversion in chromosome 1.

THE INVERSION IN CHROMOSOME 4 IS ASSOCIATED WITH A DELETION AND RETROELEMENTS.

The inversion in chromosome 4 is specific to the chimpanzee lineage as evidenced by the fact that human, gorilla, and orangutan have the same conformation in chromosome 4; only chimpanzee and bonobo have the inversion. In 2005, Kehrer-Sawatzki et al. published their detailed analysis of the DNA sequences at the breakpoints of this inversion.[3] A deletion of about 4,700 base pairs removed one of the breakpoints, so the exact location of this breakpoint is not known. The segment in human chromosome 4 that corresponds to the 4,700 base-pair deletion contains mostly nonfunctional repetitive sequences and has no genes in it, so the chimpanzee genome lost no important genetic information with this deletion. This segment may have been deleted during the inversion process, or the deletion may have happened at any point between the time of the inversion and the divergence of the chimpanzee and bonobo lineages.

Kehrer-Sawatzki et al. identified the location of the other break-point within a region rich in TA repeats (...TATATATATATA...). The region in human chromosome 14 corresponding to the 4,700 base-pair deletion also contains a TA-rich repeat region. Both regions where the breakpoints reside are also unusually rich in retroelements embedded within a longer segment that is duplicated in reverse orientation at the sites of both breakpoints. This pattern suggests that the inversion happened through recombination between two TA-rich regions that aligned when the inverted repeats aligned, in the same manner as diagrammed in figure A2.2, with subsequent deletion of one of the breakpoints.

THE BREAKPOINTS OF THE CHIMPANZEE-SPECIFIC INVERSION IN CHROMOSOME 5 CONTAIN SEQUENCES THAT RESULT IN CONFORMATIONAL CHANGES IN DNA.

The human and orangutan genomes contain the original ancestral conformation in chromosome 5. An inversion happened in chromosome 5 in the chimpanzee lineage and a translocation took place in the same chromosome in the gorilla lineage. Szamalek et al. published in 2005 their analysis of the breakpoints of this inversion.[4] Unlike the situation with chromosomes 1 and 5, there are no inverted repeats at the breakpoint sites, although there are several imperfect inverted repeats nearby that could have paired to bring the inversion breakpoints into juxtaposition.

At the breakpoints themselves are sequences that are known to cause conformational changes in DNA so that the DNA in these regions does not have the typical double-helical structure. In such regions, DNA is prone to break in both strands. Szamalek et al. proposed that inverted repeats may have caused the chromosome to loop, as in part (b) of figure A2.2, and then breaks in the region's altered DNA resulted in the inversion.

To determine if this site is prone to rearrangement in other species, Szamalek et al. compared this region of chromosome 5 with corresponding regions in the chromosomes of chicken, mouse, and rat, and found that although there were no rearrangements in these species at these exact sites, the general regions were prone to rearrangement, whereas other parts of the corresponding chromosomes were not. This suggests that these sites are predisposed to rearrangement, which resulted in an inversion in the chimpanzee lineage.

Intriguingly, the gorilla genome also contains a rearrangement in this chromosome—a translocation instead of an inversion. Translocations have only a single breakpoint in each of the two chromosomes involved. Is the single breakpoint in the gorilla translocation at or near one of the chimpanzee inversion breakpoints? It turns out that the gorilla breakpoint is completely unrelated to the chimpanzee breakpoints. It is located about ten million base pairs away from the site corresponding to the nearest inversion breakpoint in the chimpanzee chromosome. This translocation is specific to the gorilla genome and must have happened after the divergence of the gorilla lineage from the human-chimpanzee lineage.

THE DIFFERENCES BETWEEN THE HUMAN AND CHIMPANZEE VERSIONS OF CHROMOSOME 9 IN THE INVERSION REGION ARE COMPLEX.

Chromosome 9 in humans and the great apes has a long evolutionary history of inversions. Before the divergence of humans and the great apes from their shared common ancestry, an inversion took place that all humans and great apes have inherited. After the divergence of the gorilla and orangutan lineages, a second inversion occurred in the common ancestral lineage of humans and chimpanzees. Then, after the divergence of the human-chimpanzee lineages, another inversion happened in the chimpanzee lineage, which is the inversion that distinguishes chromosome 9 in the human and chimpanzee genomes. Chromosome 9 is still inversion-prone. In modern humans, some people carry newly arisen inversions in this chromosome, and some of these inversions are associated with human infertility or subfertility.[5] This chromosome is also inversion-prone in both Old World and New World monkeys.[6]

Although inversions are common in chromosome 9, the inversion that distinguishes the human and chimpanzee versions of this chromosome is unique to chimpanzees and bonobos. In 2005, Kehrer-Sawatzki et al. published their detailed analysis of the breakpoints of this inversion compared with the corresponding sequences in human chromosome 9 and found that this region has undergone complex changes in both human and chimpanzee since the divergence of their ancestral lineages.[7]

One of the breakpoints (the upper one in chimpanzee chromosome 9 in figure A2.1) aligns with its corresponding position in human chromosome 9 but cannot be exactly placed because it happened within highly repeated sequences. The other breakpoint (the lower one in chimpanzee chromosome 9 in figure A2.1) is especially interesting. It is located next to the centromere in chimpanzee chromosome 9, only twenty-three base pairs away from repetitive centromere sequences. The breakpoint itself is within a retroelement that is present in both chimpanzee and human versions of the chromosome. The centromere in the chimpanzee chromosome is out of place when compared to the human version—it is too close to the breakpoint to be explained by the inversion alone. Possibly, a second very small inversion in the chimpanzee chromosome placed the centromere near the breakpoint, but evidence collected by Kehrer-Sawatzki et al. argues against this possibility. Instead, their evidence suggests that a new centromere formed near the breakpoint after the inversion was established and that the previous centromere—the same one that functions as a centromere in the human chromosome—was inactivated in the chimpanzee chromosome. They found evidence of the original centromere exactly where it should reside, lending convincing experimental evidence to this explanation.

The inversion region differs even more between the human and chimpanzee versions because a fairly high amount of highly repetitive (and probably nonfunctional) DNA has accumulated in this region of human chromosome 9 since the divergence of the human and chimpanzee lineages, making human chromosome 9 larger than its chimpanzee counterpart. Notice the size difference between the two chromosomes in figure A2.1.

THE INVERSION IN CHROMOSOME 12
DISRUPTED A GENE AND DUPLICATED DNA.

The inversion in chimpanzee chromosome 12 is unique among the nine inversions that distinguish human and chimpanzee chromosomes in that it disrupts a gene. One of the breakpoints happened within the eleventh intron of a gene named *SLCO1B3*, disrupting its ability to function and turning it into a pseudogene. The corresponding gene in human chromosome 12 resides at the same position and is functional. Kehrer-Sawatzki et al. published their analysis of the breakpoints in 2005 and found that DNA has been duplicated at both breakpoints, probably as a result of the inversion process.[8] One of these duplications contains a fully intact copy of the *SLCO1B3* gene, which is now inverted and on the other side of the centromere as a result of the inversion. Thus, the chimpanzee genome still has a functional copy of the *SLCO1B3* gene even though the inversion disrupted the original gene.

THE BREAKPOINT OF THE CHROMOSOME 15 INVERSION RESIDES IN A TANDEM-REPEAT REGION.

As shown in figure A2.1, the chimpanzee-specific inversion in chromosome 15 has a single breakpoint. In 2003, Locke et al. published their analysis of this breakpoint.[9] They found that it resides in a region with a segment of about thirty thousand base pairs, called GLP/LCR_{15}, repeated eight times in tandem. Because the breakpoint is within a repeated region, they were unable to identify its exact site.

THE INVERSIONS IN CHROMOSOME 16 OF CHIMPANZEE AND GORILLA ARE INDEPENDENT.

Microscopic examination of human and great-ape chromosomes suggests that chimpanzees and gorillas share the same inversion in chromosome 16, whereas humans and orangutans both have the original ancestral conformation. This observation could be most easily explained if chimpanzees and gorillas were more closely related to each other than either is to humans. However, this conclusion runs counter to the resolution of the trichotomy problem (discussed in chapter 4) in which abundant evidence shows that humans and chimpanzees are more closely related to each other than either is to gorillas. This apparently shared chimpanzee-gorilla inversion was one of the lines of evidence that brought about the trichotomy problem in the first place.

In 2005, Goidts et al. examined the inversion breakpoints in both chimpanzee and gorilla chromosome 16 as compared to human chromosome 16, which lacks the inversion.[10] In the ancestral lineage common to humans, chimpanzees, and gorillas, after divergence from the lineage leading to orangutans, a segment of chromosome 1 was copied and transposed to chromosome 16. One of the inversion breakpoints lies within this transposed region. An inverted repeat of about twenty-three thousand base pairs, containing several retroelements, is located near both inversion breakpoints, suggesting that the general mechanism depicted in figure A2.2 facilitated the inversion.

Detailed examination of the breakpoints in the chimpanzee and gorilla versions of chromosome 16 reveal that both breakpoints in chimpanzee are different than those in gorilla. Although the two inversions appear to be the same when examined microscopically, DNA analysis shows that their breakpoints are different. Thus, there were two independent inversion events, one exclusively in the lineage leading to gorilla, and the other exclusively in the lineage leading to chimpanzee. This observation resolves the trichotomy discrepancy—the fact that the two inversions arose independently is consistent with the evidence that humans and chimpanzees are more closely related to each other than either is to gorillas.

THE EXACT BREAKPOINTS OF THE INVERSION IN CHROMOSOME 17 HAVE BEEN IDENTIFIED.

In 2002, Kehrer-Sawatzki et al. identified the exact breakpoints for the chimpanzee-specific inversion in the DNA sequences for chromosome 17 in humans and chimpanzees.[11] As shown in figure A2.3, the five base-pair sequence GGGGT is present in both breakpoints. This sequence is a target of certain enzymes called endonucleases that cut DNA, suggesting that DNA

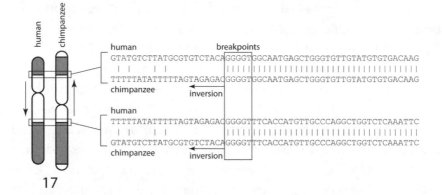

Figure A2.3. Comparison of DNA sequences at the inversion breakpoints in the human and chimpanzee versions of chromosome 17. Only one strand of each DNA molecule is shown. Vertical lines denote identical nucleotides in the human and chimpanzee sequences.

cleavage by an endonuclease may have been responsible for the breaks that facilitated the inversion.

THE HUMAN-SPECIFIC INVERSION IN CHROMOSOME 18 PROBABLY AROSE THROUGH DUPLICATION OF DNA FOLLOWED BY RECOMBINATION.

In 2004, Goidts et al. described in detail the features of the inversion in human chromosome 18.[12] This inversion is found exclusively in humans, and the events that precipitated its occurrence were also specific to humans. The breakpoints of the inversion are within inverted repeats found on either side of the centromere in the human chromosome. Each repeat is approximately nineteen thousand base pairs long. The presence of these repeats at the breakpoints of the inversion is strong evidence that this is yet another example of an inversion that arose through the recombination model depicted in figure A2.2. The recombination could have happened anywhere within the paired nineteen thousand base-pair repeats.

Interestingly, part of a gene called *ROCK1* extends into one of the repeats, and that part of the *ROCK1* gene is present in the other repeat as a pseudogene fragment (see figure A2.4). It is possible that the recombination took place within the piece of the *ROCK1* gene in the repeat. However, even if this happened, recombination between the gene and a fragment of it simply attaches the same sequence onto the gene, so the gene remains intact. Thus, regardless of where the exact breakpoint is, whether in the gene or not, the *ROCK1* gene is not disrupted by the inversion and it functions normally.

As shown in figure A2.4, only one copy of the nineteen thousand base-pair segment is present in chimpanzee chromosome 18, evidence that the repeat arose by duplication of this segment exclusively in the human lineage after the human and

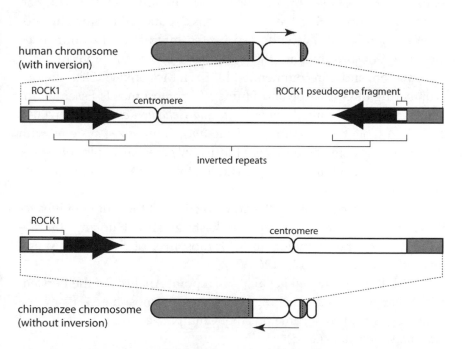

Figure A2.4. Comparison of the structures in human and chimpanzee versions of chromosome 18. The *ROCK1* gene is present at the same position in both species, but the inverted repeat carrying a pseudogene fragment of the *ROCK1* gene is unique to humans.

chimpanzee lineages diverged. Chimpanzee chromosome 18 also has the *ROCK1* gene in the same position as in humans but lacks the pseudogene fragment.

NOTES

1. All articles cited in this appendix by Kehrer-Sawatzki et al., Szamalek et al., and Goidts et al. represent research conducted at the University of Ulm under the direction of Dr. Hildegard Kehrer-Sawatzki with collaboration from scientists at the University of Ulm and other institutions. These authors consistently refer to chimpanzee chromosomes by the traditional numbering system in which chimpanzee chromosomes are numbered by size from largest to smallest with chromosome 1 as the largest. I have instead designated chimpanzee chromosomes by numbers corresponding to their human counterparts (including the designations 2A and 2B for the chromosome 2 counterparts) as endorsed by the International Chimpanzee Genome Consortium and the National Center for Biotechnology Information, and as recommended by E. H. McConkey, "Orthologous Numbering of Great Ape and Human Chromosomes Is Essential for Comparative Genomics," *Cytogenetic and Genome Research* 105 (2004): 157–58. A table for comparison of the two chimpanzee systems with the human system can be viewed at http://www.ncbi.nlm.nih.gov/genome/guide/chimp/chimpchrtable.html (accessed January 20, 2007).

2. J. M. Szamalek et al., "Characterization of the Human Lineage-Specific Pericentric Inversion that Distinguishes Human Chromosome 1 from the Homologous Chromosomes of the Great Apes," *Human Genetics* (2006) 120: 126–38.

3. H. Kehrer-Sawatzki et al., "Breakpoint Analysis of the Pericentric Inversion Distinguishing Human Chromosome 4 from the Homologous Chromosome in the Chimpanzee (*Pan troglodytes*)," *Human Mutation* 25 (2005): 45–55.

4. J. M. Szamalek et al., "Molecular Characterisation of the Pericentric Inversion that Distinguishes Human Chromosome 5 from the

Homologous Chimpanzee Chromosome," *Human Genetics* 117 (2005): 168–76.

5. G. Collodel et al., "TEM, FISH and Molecular Studies in Infertile Men with Pericentric Inversion of Chromosome 9," *Andrologia* 38 (2006): 122–7; M. Srebniak et al., "Subfertile Couple with inv(2), inv(9) and 16qh+," *Journal of Applied Genetics* 45 (2004): 477–79; I. P. Davalos et al., "Inv(9)(p24q13) in Three Sterile Brothers," *Annals of Genetics* 2000 43 (2000): 51–54; I. Sasagawa et al., "Pericentric Inversion of Chromosome 9 in Infertile Men," *International Urology and Nephrology* 30 (1998): 203–207.

6. G. Montefalcone et al., "Centromere Repositioning," *Genome Research* 9 (1999): 1184–88.

7. H. Kehrer-Sawatzki et al., "Molecular Characterization of the Pericentric Inversion of Chimpanzee Chromosome 11 Homologous to Human Chromosome 9," *Genomics* 85 (2005): 542–50.

8. H. Kehrer-Sawatzki et al., "Breakpoint Analysis of the Pericentric Inversion between Chimpanzee Chromosome 10 and the Homologous Chromosome 12 in Humans," *Cytogenetic and Genome Research* 108 (2005): 91–97.

9. D. P. Locke et al., "Refinement of a Chimpanzee Pericentric Inversion Breakpoint to a Segmental Duplication Cluster," *Genome Biology* 4 (2003): R50.

10. V. Goidts et al., "Independent Intrachromosomal Recombination Events Underlie the Pericentric Inversions of Chimpanzee and Gorilla Chromosomes Homologous to Human Chromosome 16," *Genome Research* 15 (2005): 1232–42.

11. H. Kehrer-Sawatzki et al., "Molecular Characterization of the Pericentric Inversion that Causes Differences between Chimpanzee Chromosome 19 and Human Chromosome 17," *American Journal of Human Genetics* 71 (2002): 375–88.

12. V. Goidts et al., "Segmental Duplication Associated with the Human-Specific Inversion of Chromosome 18: A Further Example of the Impact of Segmental Duplications on Karyotype and Genome Evolution in Primates," *Human Genetics* 115 (2004): 116–22.

Appendix 3

FROM DARWIN TO THE HUMAN GENOME

A Brief History

Since the dawn of human civilization, people have wondered about their origins. Who were our remote ancestors? Where did they live? Why do we resemble other mammals in so many respects? Why do children look so much like their parents and siblings? For millennia, the answers to these and similar questions were shrouded in mystery and myth. By the mid-nineteenth century, science had offered some clues but no solid answers. Then three paths of discovery began, almost simultaneously, which later would converge to lead us where we are now. This appendix offers a brief history of the scientists who laid the foundation that made possible the discoveries presented in this book.

The first path began with Charles Darwin's publication of the *Origin of Species* in 1859. The second started when Gregor Mendel deduced the fundamental principles of inheritance, which he published in 1866. Three years later, the third originated with Friedrich Miescher's discovery of DNA in 1869. The first two paths intermingled from time to time then fully converged in the 1930s. The third path, which added molecular evidence to classical theories, began to join the other two in 1944 and by the 1970s was a full-fledged component.

DARWIN AND MENDEL WERE CONTEMPORARIES, BUT THEIR THEORIES REMAINED SEPARATE DURING THEIR LIFETIMES AND LONG AFTER.

Darwin's *Origin of Species* was nothing short of a scientific revolution from the day of its publication. Although evolutionary ideas had been a part of science for some time, Darwin's book brought together evidence from a variety of sources—animal and plant breeding, geology, biogeography, comparative anatomy, and others—to explain how all of life was genetically related and how the diverse array of species arose through gradual evolution from common ancestry by natural selection over very long periods of time.

In 1858, Darwin discovered that Alfred Russell Wallace had independently come up with an idea of natural selection that was so close to his own that Darwin referred to it as a "striking coincidence," and that "if Wallace had my M.S. sketch written out in 1842 he could not have made a better short abstract!"[1] This discovery motivated Darwin to frantically finish the *Origin of Species* and rush it to publication the following year.

An absolute requirement of Darwin's theory was inheritance of variations, a requirement Darwin often reiterated. Unfortunately, Darwin's understanding of heredity was incomplete, and in most respects incorrect. Unbeknownst to Darwin, during that fast-paced year of 1858, Gregor Mendel was in the midst of his monumental experiments that would lead him to deduce the fundamental principles of inheritance that remain essentially unchanged today.

Mendel was a friar in the Augustinian St. Thomas monastery in the city of Brno in what then was part of Austria but is now in the Czech Republic. In 1848, the monastery abbot sent Mendel to the University of Vienna for two years to study natural science so that he could be certified as a science teacher. Among Mendel's most influential university teachers was his botany professor, Franz Unger, who openly taught that species had evolved over eons of time and were still evolving. Such a view was unusual for the time; Darwin's *Origin of Species* had not yet been published. In an article Unger wrote for the local newspaper while Mendel was his student, he foreshadowed Mendel's work in these words: "Who can deny that new combinations arise out of this permutation of vegetation, always reducible to certain law-combinations, which emancipate themselves from the preceding characteristics of the species and appear as a new species?"[2]

That Mendel viewed his work as pertinent to evolution is clear from the introductory section of his classic paper, where he proposed that his experiments were "the one correct way of finally reaching the solution to a question whose significance for the evolutionary history of organic forms must not be underestimated."[3]

Even though *Origin of Species* had yet to be published, a clash between evolution and religion was already under way, with Mendel in the midst of it. Unger's teachings on the mutability of species drew heated criticism from the Viennese clergy, some of whom vigorously fought to have him dismissed from his university post. Among the most vocal was Dr. Sebastian Brunner, who labeled Unger as "a man who openly denied the Creation and the Creator" and as one of the "professors at so-called Catholic Universities [who] deliver lectures on really beastly theories for years on end."[4] Although Mendel was a devout member of the clergy, he clearly had great respect for Unger. Years later Mendel sent him what is now one of the few surviving reprints of his classic paper.[5]

Mendel painstakingly hybridized pea plants and studied

patterns of inheritance in thousands of their offspring over a period of eight years. He finished his experiments in 1863 but did not publish them until 1866. By then he had purchased his own copy of *Origin of Species*, a German translation published in 1863, in which he made marginal notes. It is apparent from these notes, and from passages in Mendel's paper, that *Origin of Species* influenced his writing, although the dates make it clear that Darwin's book could not have inspired or affected the experiments themselves.[6]

The immediate fate of Mendel's work is one of the greatest misfortunes in the history of science. Mendel presented his work in two lectures to the local natural history society and published it in the society's journal, which was held in libraries throughout Europe and even a few in the United States. Regrettably, no one, not even Mendel himself, recognized the importance of his work.

Even more unfortunate was the fact that Darwin never knew of Mendel's experiments. Numerous books, articles, and Web sites claim that Darwin had a reprint of Mendel's paper, but the pages were uncut and thus unread. The truth, however, is a bit less dramatic. There is no evidence that Darwin had a copy of Mendel's paper. Instead, he owned a book by Focke summarizing hundreds of hybridization experiments, including brief references to Mendel's.[7] The pages in Darwin's copy of Focke's book with references to Mendel's work were uncut.[8]

In 1868, Darwin published his own hypothesis of heredity, which he called *pangenesis*, the notion that each organ in the body releases hereditary particles (which Darwin called gemmules) into the blood or plant sap. The gemmules supposedly coalesced into the egg cells and sperm cells or pollen grains, which transmitted them to a new organism at the time of conception. Darwin wrote, "I give my well-abused hypothesis of Pangenesis. An unverified hypothesis is of little or no value; but if anyone should hereafter be led to make observations by which some such hypothesis could be established, I shall have done good service, as an astonishing number of isolated facts can be thus connected together and rendered intelligible."[9]

Mendel happened across Darwin's hypothesis of pangenesis while reading a German translation of Darwin's *Plants and Animals under Domestication* and recognized that this was in error. In the margin by one of the pages describing pangenesis, Mendel wrote "*sich einem Eindrucke ohne Reflexion hingeben* [to succumb to an impression without giving the matter proper thought]."[10]

With Darwin's encouragement, his cousin, Francis Galton, tested the hypothesis of pangenesis experimentally. Galton presumed that if the gemmules are in the blood of rabbits, then blood transfused from one rabbit to another should introduce the donor rabbit's gemmules into the recipient rabbit, and those gemmules should then be transmitted to the recipient rabbit's offspring. Part of his enthusiasm for these experiments came from their potential for practical use. If, according to the hypothesis of pangenesis, gemmules circulated in the blood, then animal breeders could introduce inherited traits from one breed to another simply through blood transfusions.

From 1869 to 1872, Galton conducted many transfusion experiments, in some cases attempting to displace as much as half of the recipient rabbit's blood with the donor's. But contrary to his expectations, he failed to find transmission of the characteristics of donor rabbits in the progeny of recipient rabbits. Referring to her husband and Galton, Darwin's wife lamented in a letter to her daughter, "F. Galton's experiments about rabbits are failing, which is a dreadful disappointment to them both."[11] Thus, not long after he proposed it, Darwin found that his hypothesis of pangenesis as the mechanism of heredity was incorrect.

In 1870, Mendel was elected abbot of his monastery and spent the rest of his life embroiled in bitter dispute over taxation of church property. His scientific career effectively ended the following year with the publication of his last scientific paper, a hypothesis on how tornadoes form.

Some authors claim that as a priest Mendel was an opponent of Darwinism and a proponent of special creation.[12] There is little foundation for such views. Historical evidence reveals that, although Mendel did not always agree with Darwin's ideas on heredity and botany, he perceived his own work as directly related to the process of evolution and the formation of new species.[13]

Most of Mendel's writings are gone, probably burned after his death. Those that remain suggest that he kept his religious and scientific views separate. His scientific writings portray nothing religious, only the astute mind of a meticulous and knowledgeable scientist. His surviving religious writings, and recollections of those who knew him, reveal a liberal-minded, obstinate, and progressive priest who nonetheless remained devoted to his faith throughout his life in spite of disputes with his superiors. He died in 1884, highly respected as a priest, teacher, and friend to the poor—but virtually unknown as a scientist.

Darwin's fate was the opposite. He died in 1882, famous throughout the intellectual world for his contributions to science. *Origin of Species* was in its sixth edition, and Darwin had published other highly regarded and influential books, several of which remain classics today, among them *The Descent of Man, and Selection in Relation to Sex*, a book that focused on human evolution.

Was there a connection between Mendel and Darwin? The answer is yes, but it was entirely one-way. Mendel knew of Darwin and read his books. Darwin knew nothing of Mendel.

THE STORY OF DNA BEGINS IN 1869.

Mendel proposed in his 1866 paper that there were "potentially formative elements" that determined inherited characteristics.

He did not attempt to speculate on the substance of those elements, and nothing in his experiments could have led him to the answer. However, about 570 kilometers to the west in Tübingen, Germany, Friedrich Miescher, a twenty-five-year-old chemist who had recently completed his doctoral degree, discovered the substance of those elements—DNA—in 1869. In one of his experiments, Miescher isolated a substance from cells that he found to have an unusually high phosphorus content. He recognized that the substance he had discovered was unique and named it *nuclein* because it was abundant in the cell nucleus.

Miescher continued to study nuclein until his untimely death at age fifty-one in 1895, leaving much of his work unpublished. So devoted was Miescher to his work that when he failed to arrive at his wedding, his friends searched for him and found him finishing experiments in the laboratory.

Nuclein was eventually found to contain some protein as well as a mixture of DNA and RNA. In 1889, Richard Altmann, one of Miescher's successors, named this mixture *nucleic acid*, to distinguish it from the protein in nuclein. Although the term *nuclein* is no longer used, *nucleic acid* remains in use today to describe both DNA and RNA.

Albrecht Kossel continued much of the work with nucleic acid, which was usually a purified mixture of varying amounts of DNA and RNA. After exhaustive chemical analysis over a period of years, he identified one by one the chemical constituents of DNA. By 1894, he had identified all four of the bases in DNA, guanine, adenine, cytosine, and thymine (G, A, C, and T). One of those bases, guanine (G), had already been described chemically; Kossel described and named the other three. The last of the bases that Kossel discovered was thymine (T), and he named it for the thymus gland because that is where he found it.

For his efforts, Kossel received a Nobel Prize in 1910. Unfortunately, Kossel left much of his work unpublished in papers stored at the University of Heidelberg. When Allied soldiers invaded Heidelberg during World War II, they burned Kossel's papers to keep themselves warm.

There was little reason during the late nineteenth century for anyone to suspect that DNA might be the substance of heredity. Instead, most believed that it was a reservoir of phosphorus for use by the cell, a view that prevailed for decades. In one brief instance, Edmund Wilson, a famous American cell biologist at Columbia University, hypothesized that DNA might be the hereditary material when he wrote, "There is considerable ground for the hypothesis that in a chemical sense this substance [nucleic acid] is the most essential nuclear element handed on from cell to cell, whether by cell-division or fertilization."[14] Wilson's hypothesis was soon forgotten, and more than forty years would pass before DNA was recognized as the substance of heredity.

MENDEL'S WORK WAS REDISCOVERED IN 1900, INITIATING THE SCIENCE OF GENETICS AND RELEGATING DARWIN'S WORK TO A SECONDARY POSITION.

After lying mostly dormant in Mendel's article for thirty-four years, a correct view of inheritance took the world of science by storm with the rediscovery of Mendel's theory in the year 1900. In the late 1890s, three botanists, Carl Correns, Hugo de Vries, and Erich Tschermak von Seysenegg, independently conducted the experiments that were destined to bring widespread recognition to Mendel's work. All three worked with plants and all three observed the same patterns as Mendel, albeit with fewer and smaller experiments, and all three independently arrived at the same conclusions as Mendel. They separately published their studies in 1900, initiating the science of genetics with the dawn of a new century.[15]

Some scientists embraced Mendel's theory; others claimed it

was a rare exception to the general pattern of inheritance. Unfortunately, many saw it as contrary to Darwinism. British naturalist William Bateson was the most passionate of the early Mendelians. Although not one of the three rediscoverers, he immediately embraced Mendelism, christened the new science *genetics* (a word he coined), and seized the opportunity to preach it to the world. Through exhaustive experimentation, Bateson and his followers conclusively showed that Mendelian inheritance was indeed the rule and not the exception.

Bateson was not especially supportive of Darwinism, although he did not dismiss Darwin's views altogether. However, the Mendel-Darwin dichotomy, as some perceived it, ended up favoring Mendel as scientists experimentally confirmed Mendelian inheritance in one species after another, including humans. Almost three decades would elapse before population biologists would demonstrate that Mendelian genetics and Darwinian evolution are mutually supportive of each other.

A SERIES OF DISCOVERIES EMERGE FROM MORGAN'S "FLY ROOM."

The year 1910 was a major milestone. Thomas Hunt Morgan, a young professor at Columbia University in New York, had chosen to work with the common fruit fly, *Drosophila melanogaster*, as a model organism for studying evolution. At first, he was skeptical of Mendelism, but his repeated observations of inheritance in fruit flies quickly converted him into a staunch Mendelian. In 1910, Morgan observed one fly that was to change his thinking, his career, and the science of genetics. The fly was a male with white rather than the typical brick-red eyes. He allowed the male fly to mate with a red-eyed female and then followed the patterns of inheritance in the progeny. They were somewhat Mendelian but not quite. He found that white eyes appeared more often, but not exclusively, in males, in predictable ratios. It dawned on him that inheritance of the gene responsible for the white eyes was entirely associated with the inheritance of the X chromosome.

Morgan and his students soon found other genes that behaved in the same way. The inevitable conclusion soon was obvious: *genes are located on chromosomes*. Three of Morgan's students, Alfred Sturtevant, Calvin Bridges, and Hermann Muller, worked closely with Morgan in his "fly room." All three would be among history's most influential biologists. One of the first to make a major discovery was Sturtevant. By 1911, Morgan and his students had discovered several genes on the X chromosome of fruit flies but wondered how they were arranged on the chromosome. Sturtevant was an undergraduate student at the time, and it dawned on him that he might be able to determine the relative locations of the genes on the X chromosome simply by calculating how often offspring carried mutated versions of each gene. One day he gathered up the lab notebooks, took them home, and, as he put it, "spent most of the night (to the neglect of my undergraduate homework) in producing the first chromosome map."[16] Sturtevant's gene mapping method would eighty years later be an essential key to assembling the sequence of the human genome.

The 1920s was the age of the chromosome. Lilian Morgan and Calvin Bridges discovered the mechanism of sex determination with X and Y chromosomes in the fruit fly.[17] Harriet Creighton and Barbara McClintock used a brilliant synthesis of microscopic examination of chromosomes with classical genetics to show how genes are physically located on chromosomes.[18] Sturtevant and Bridges discovered inversions and translocations, respectively, in chromosomes of the fruit fly.[19] The stage was now set for the emergence of modern evolutionary genetics.

THE MODERN SYNTHESIS ARRIVES IN THE 1930s.

During the 1920s and culminating in the 1930s, four geneticists, J. B. S. Haldane, Ronald A. Fisher, Sewall Wright, and Theodosius Dobzhansky, reinterpreted Darwin's reasoning in the light of Mendelian and chromosomal genetics, establishing a new version of Darwinian thought, now called *neo-Darwinism*, explained in Mendelian genetic terms. This synthesis of Mendelism and Darwinism is now known as *the modern synthesis*. Each of these four scientists laid one of the cornerstones in the foundation of the modern synthesis and brought Darwin's work back into prominence.

In 1924, Haldane laid the first cornerstone when he began publishing a series of influential articles in which he derived equations that mathematically explained natural selection in Mendelian terms. His mathematical models were later shown to be correct when they were tested experimentally in animal and plant populations. Scientists today continue to use the mathematical tools he developed.

In 1930, Fisher published *The Genetical Theory of Natural Selection*, a book that, according to science historian J. H. Bennett, is "celebrated as the first major work to provide a synthesis of Darwinian selection and Mendelian genetics."[20] Throughout the book, Fisher took Darwin's reasoning on the role of natural selection and mathematically reinterpreted it in the light of Mendelian genetics. As he finished each chapter, he sent the text to Darwin's son Leonard for comments on how well he represented Charles Darwin's views.

As is so often the case for a work destined to become a classic, Fisher's book did not immediately gain widespread recognition; it took seventeen years for the first printing of fifteen hundred copies to sell out. By way of consolation, Leonard Darwin wrote to Fisher, "It will be slowly recognized as a very important contribution to the subject. But I am afraid that it will be slow, because so few will really grasp all that it means. You must not, therefore, be disappointed at the reception which it receives, but trust to the ultimate results."[21] These words rang true; Fisher's book became the second cornerstone in the foundation and remains a pivotal work in evolutionary biology.

In 1931, Wright published a long and detailed article in the journal *Genetics* titled "Evolution in Mendelian Populations" that would be the third cornerstone of the modern synthesis.[22] In this article, he introduced what he called the "three-phase shifting balance theory" of evolution. This hypothesis proposed that natural selection was one of three major forces in evolution, the other two being random fluctuations in genetic composition in small populations (a phenomenon known as *genetic drift*), and the effect of migration on dispersal of genes.

Wright's theory of genetic drift explains an extremely important phenomenon. It predicts that over many generations nearly every mutation or chromosomal rearrangement will eventually suffer one of two fates: it will either spread throughout the species until every member of the species has it (a phenomenon known as *fixation*), or it will disappear entirely from the species. Natural selection can help the process along by fixing favorable mutations and eliminating unfavorable ones, but even selectively neutral mutations or chromosomal rearrangements will eventually be either fixed or lost in the species. Experiments in a variety of species have confirmed that genetic drift follows the very pattern that Wright mathematically predicted. This theory explains why every human being has a fused chromosome 2, the same inversions on chromosomes 1 and 18, and, for the most part, the same transposable elements and pseudogenes. Genetic variations (including mutations, transposable elements, pseudogenes, and chromosome rearrangements) that differ among people alive today arose in relatively recent times on an evolutionary timescale; the older ones have become fixed or lost. The same is true for all other living species.

After his retirement in 1954, Wright worked for more than

two decades writing a four-volume masterpiece titled *Evolution and the Genetics of Populations*, published between 1968 and 1978, that covered in detail the history and theory of evolutionary genetics.[23]

Haldane, Fisher, and Wright's contributions to evolutionary genetics were mostly theoretical inferences based on their applications of mathematics to Darwinian selection and Mendelian inheritance. Dobzhansky's contribution was both theoretical and experimental. In 1933, he published an article on the evolution of chromosomal inversions in fruit flies. Then, over a period of forty years, from 1936 to 1976, Dobzhansky and Sturtevant published a series of forty-three papers collectively titled "Genetics of Natural Populations," in which they documented through experimental analysis the evolution of natural populations of fruit flies, much of it based on inversions.

In 1937, Dobzhansky published the fourth cornerstone of the modern synthesis, a book titled *Genetics and the Origin of Species*. In this book, he integrated the results of more than three decades of laboratory and field research with the theoretical principles expounded by Haldane, Fisher, Wright, and others.

STUDIES ON DNA WERE INDEPENDENT OF GENETICS AND EVOLUTION UNTIL 1944.

It is often said (incorrectly) that James Watson and Francis Crick discovered DNA. On the contrary, they deduced just the final component of DNA's structure—how two strands of DNA wrap around each other to form a double helix—based on a wealth of information collected over a period of eighty-four years. By the first decade of the 1900s, the discoveries of Miescher and Kossel had shown the elemental composition of DNA and that DNA has four bases. But no one yet knew what all of the molecule's components were or how they fit together.

Phoebus Levene, a Russian immigrant to the United States, had studied in Kossel's laboratory in Germany, an experience that directed his interests toward DNA. Kossel had found that nucleic acid contains a sugar but he was not able to identify it. In 1909, Levene purified and described ribose, a sugar found in RNA rather than DNA. Twenty years later, in 1929, Levene identified the final component of DNA: the sugar deoxyribose, which is missing an oxygen atom when compared to ribose, hence its name *deoxyribose*.

The three components of DNA, the sugar deoxyribose, the bases (T, C, A, and G), and phosphate, were now known. The next challenge was determining how they are assembled in a DNA molecule. Through exhaustive experimentation, Levene discovered that a single base is attached to each deoxyribose sugar and that one or more phosphate groups are also attached to the sugar to form what is called a *nucleotide*.

In 1935, Levene proposed that four nucleotides in DNA were connected to one another as a chain, each with one of the four bases. Levene's model was called the *tetranucleotide hypothesis* because he hypothesized that each chain consisted of only four nucleotides. Some of Levene's experiments suggested to him that the relative quantities of T, C, A, and G in DNA are always equal, a conclusion that would later be disproved but one that seemed to support his hypothesis.

Although most DNA molecules contain thousands to hundreds of millions of nucleotides, Levene's model correctly showed how nucleotides were attached to one another in DNA. In 1938, Levene published the results of experiments showing that DNA molecules are substantially larger than a tetranucleotide. However, he unfortunately surmised that these large DNA molecules are simply chains of repeating tetranucleotides. Levene died in 1940, unaware that DNA is the genetic material and that his work would be one of the most important pieces that scientists would later use to discover the structure of DNA.

Levene's tetranucleotide hypothesis was a major step in the history of DNA research and it revealed Levene's remarkable insight. It also had an unfortunate negative effect. By the 1930s, from the work of McClintock, Creighton, Sturtevant, Bridges, Muller, Morgan, and others, biologists knew that chromosomes carry the genetic material. They also knew that chromosomes are composed of DNA and protein, so either DNA or protein had to be the genetic material. The tetranucleotide hypothesis suggested that DNA was a repeating tetranucleotide with little diversity in its structure—and diversity in the genetic material is essential for it to encode the tremendous inherited diversity in life. Proteins consist of twenty different amino acids arranged in a variety of combinations. Thus, protein had the necessary diversity and seemed to most scientists to be the logical candidate for the genetic material.

In the meantime, Frederick Griffith, a reclusive and dedicated British medical scientist, began a search that would eventually prove DNA, rather than protein, to be the genetic material. Griffith had noticed that patients with pneumonia were often infected with different types (called *strains*) of the same species of bacteria. His observations led him to conclude that one strain increased while another decreased when patients contracted pneumonia. The one that increased was the *virulent* (disease-causing) strain, and the one that decreased was the *avirulent* (harmless) strain. Suspecting that one strain might convert a second strain into the first, he began a series of experiments that ultimately revealed that some substance transferred from the virulent strain converted the avirulent strain into the virulent one.

He called this substance the *transforming principle*, but he did not live to discover what the substance was. Griffith was very reclusive and dedicated, working day and night in his laboratory. Tragically, in 1941 a German bomb struck his laboratory during the blitz of London, and he met his death there, unaware that his work would soon become famous.

Oswald Avery was an established bacteriologist working at the Rockefeller Institute in New York City in the early 1940s. When he read Griffith's 1928 paper on bacterial transformation, he realized that the transforming principle must be the substance of heredity and he quickly changed his research to try to identify the substance. Workers in his laboratory began the tedious work of purifying various substances from bacterial cells to see which could transform avirulent cells into virulent cells. One of those workers, Lionell Alloway, isolated a stringy whitish substance that transformed cells at the highest rate of any material tested. Thus, Avery had available to him a purified substance that consisted predominantly, if not entirely, of Griffith's transforming principle.

Avery and two collaborators in his laboratory, Colin MacLeod and Maclyn McCarty, determined the relative amounts of elements present in the substance that Alloway had purified and found that its composition matched that of DNA. They treated the substance with protease, an enzyme that destroys proteins, and with ribonuclease, an enzyme that destroys RNA, and discovered in both cases that the substance still transformed cells as efficiently as it did before treatment. However, when they treated the substance with deoxyribonuclease, an enzyme that destroys DNA, the substance no longer transformed R cells into S cells. All of the evidence from their experiments pointed to DNA as Griffith's transforming principle. When the transformed cells divided, their progeny were all virulent cells, and only virulent cells were produced after many generations of cell division. This observation showed that the transforming principle must be inherited. Therefore, DNA must be the genetic material of these bacterial cells.

Seventy-five years after Miescher discovered DNA, someone finally had identified it as the genetic material of an organism. The parallel stories of DNA, inheritance, and evolution began to converge, and the age of molecular evolutionary genetics was beginning to dawn.

THE DOUBLE-HELICAL STRUCTURE OF DNA IS DISCOVERED.

The experiments of Avery, McLeod, and McCarty had demonstrated that DNA is the hereditary material in a bacterial species. Because patterns of inheritance are the same in most species, it made sense that the genetic material was also the same among all species. Thus, the discovery that DNA is the genetic material of a bacterium suggested that it is also the genetic material of other organisms. However, such a conclusion required experimental demonstration rather than simple conjecture.

Bacterial viruses are nature's simplest genetic entities. Most of them consist of just two substances: DNA and protein. Therefore, either the DNA or the protein must be the virus's genetic material. When one of these viruses infects a cell, it injects part of itself into the cell where it then replicates to make many copies of itself. Alfred Hershey and Martha Chase surmised that whatever was injected into the cell, the DNA or the protein, must be the virus's genetic material. Through a series of experiments in which they radioactively labeled DNA and protein separately, they found that DNA was the substance injected into the cells. Therefore, DNA must be the hereditary material of the viruses they studied. They published their work in 1952.[24] However, DNA's structure had still not been determined.

The experiments of Erwin Chargaff were among the most important of several landmark studies that paved the way toward the discovery of DNA's structure. When Chargaff read Avery, MacLeod, and McCarty's 1944 paper on DNA as the transforming principle, he recognized that DNA must be the genetic material and he changed his research to focus on DNA. While studying the four bases in DNA, he noticed a peculiar pattern. Levene had reported that the four bases, T, C, A, and G, were present in equal proportions in DNA. However, Chargaff's data showed that the relative proportions of the bases varied among different species. There was, however, a consistent pattern in the variation. In all species he studied, the proportions of T and A were equal, and the proportions of G and C were equal. For example, Chargaff found that human DNA contains 20 percent T, 20 percent A, 30 percent C, and 30 percent G. He expanded his research to include DNA from more species and continued to find a consistent pattern of T = A and C = G. In a 1951 paper, he wrote, "As the number of examples of such regularity increases, the question will become pertinent whether it is merely accidental or whether it is an expression of certain structural principles."[25] The latter option would prove to be the correct answer, and that answer was only two years away.

The experiments of Avery, McLeod, and McCarty, and those of Hershey and Chase, gave the world strong evidence that DNA is the genetic material of at least some species, and perhaps all species. In the early 1950s, however, few scientists were excited about this possibility. James Watson, a young American studying in Europe, was one of the few who enthusiastically embraced the idea that DNA might be the universal genetic material and he suspected that the discovery of DNA's molecular structure would be one of the greatest achievements in the history of science.

By then, much about the structure of DNA was already known. Before 1900, Kossel had revealed the structures of all four bases, and in 1929 Levene found the correct structure of nucleotides. Levene also showed how the nucleotides were connected in a chain in his 1935 version of the tetranucleotide hypothesis. By the early 1950s, Alexander Todd, a distinguished biochemist at Cambridge University, had shown that Levene's basic model was correct but that a strand of DNA contained many more than four nucleotides, and that the nucleotides could be assembled in any order, rather than in groups of four as Levene had proposed. Thus, the challenge to scientists in the 1950s was to discover how one, two, three, or more strands of DNA assembled themselves to form a DNA molecule.

Watson went to Copenhagen as a postdoctoral fellow to work on bacterial viruses, but his experience there was not good. While at a meeting in Naples, he attended a presentation by the British scientist Maurice Wilkins on DNA and immediately became enamored with the possibility of studying the structure of DNA. He enthusiastically approached Wilkins after the speech to discuss DNA, but according to Wilkins, Watson "was a bit of a puzzle to me, I didn't quite know what to make of him."[26] Watson ended up securing a temporary research position at Cambridge University in England, where he was assigned to share an office with Francis Crick, an Englishman twelve years his senior.

Watson at first felt intimidated by Crick. Even though Watson already had a PhD, and Crick did not, Watson was still quite young at only twenty-three years of age. To make matters worse, Crick was much more adept at physics and mathematics than was Watson. However, within a short time, the two became fast friends and their lunchtime discussions inevitably gravitated to DNA.

Neither Watson nor Crick was assigned to work with DNA. Nonetheless, it became practically their sole interest. Many scholars would later hail their discovery of the structure of DNA as the greatest discovery of the twentieth century. However important it was, it was also a very unconventional discovery. Most scientists collect data from their own experiments, then interpret their data in the light of experiments published by others. Watson and Crick based their discovery entirely on the experimental data of others. Their accomplishment was a consequence of a serendipitous combination of several factors: both were enthusiastic and tenacious in their struggle to decipher the structure of DNA, their areas of expertise complemented one another, they had the genius to harmonize a vast array of diverse information, and they were lucky to be in the right place at the right time.

Perhaps their best fortune was their access to Maurice Wilkins, who was Crick's friend and whom Watson had met in Naples. Wilkins was working with his assistant, Raymond Gosling, fifty miles away from Cambridge at King's College in London on *x-ray diffraction* of DNA, a procedure that bounces x-ray beams off of molecules as a way of determining their structure. Rosalind Franklin, a physicist with a superb talent for x-ray diffraction, joined Wilkins and Gosling at King's in 1951. A misunderstanding about who was in charge of the DNA project created immediate friction between Franklin and Wilkins that was never fully resolved. While Wilkins was away on a foreign trip, Franklin and Gosling obtained some of the clearest and most informative x-ray diffraction images of DNA.

Franklin was a devoted and methodical worker, not one to rush to hasty conclusions. Thus, it was with some resentment that she, along with Wilkins, Gosling, and others from the King's lab, traveled by train to Cambridge to see a model of DNA that Watson and Crick had constructed based in part on the information Franklin had presented in her seminar. The model contained three strands of DNA wound around each other with the phosphate groups on the inside of the helix and the bases on the outside. Unfortunately, Watson and Crick had based their model on a simple but significant error, which Franklin easily recognized, and she quickly dismissed the model as incorrect. With some embarrassment, Watson later recalled Franklin and Gosling asserting that "their future course of action would be unaffected by their fifty-mile excursion into adolescent blather."[27]

This was not the only incorrect model that Watson and Crick would construct. In one model they paired like bases (T–T, C–C, G–G, and A–A) in a double helix, a model that also failed to conform with Franklin and Gosling's data. Of this model, Watson wrote, "For two hours I happily lay awake with pairs of adenine residues whirling in front of my closed eyes. Only for a brief moment did the fear shoot through me that an idea this good could be wrong."[28]

Watson and Crick had difficulties determining how the nucleotides in two or more strands of DNA interacted with one another. Crick knew a young man named John Griffith, a nephew of Frederick Griffith who discovered transformation in bacteria. The younger Griffith had devised a way for the bases in DNA to pair with each other, and he briefly mentioned his idea to Crick, who paid it little attention at the time. However, in 1952, Chargaff visited Cambridge and spoke to Watson and Crick about his publications of the T = A, C = G relationships in DNA, which neither of them had read. Crick suddenly remembered his conversation with Griffith, and (in his words), "The effect was electric. That is why I remember it. I suddenly thought 'Why my God, if you have complementary pairing you are bound to get a 1:1 ratio.' . . . And to my astonishment the pairs that Griffith said were the pairs that Chargaff said."[29]

Ultimately, Watson and Crick constructed a double-stranded helical model of DNA that met the requirements of the experimental data obtained by Franklin, Gosling, and Wilkins, and it also explained Chargaff's observations of the T = A and C = G relationships. Lastly, it adhered to all of the rules of chemistry.

When Wilkins learned of the model, he responded, "I think you're a couple of rogues but you may well have something. I like the idea."[30] Franklin agreed that the model was correct when she learned of it. Watson and Crick published their model in a short article in the journal *Nature*, which is now one of the most famous articles ever published.[31] Two articles immediately followed Watson and Crick's in the same issue of *Nature*, one by Wilkins, Stokes, and Wilson, and another by Franklin and Gosling, both on x-ray diffraction patterns of DNA.[32]

Watson, Crick, and Wilkins shared the 1962 Nobel Prize for their discovery of the structure of DNA. Most agree that Franklin deserved such recognition as well. However, just as she was beginning a promising and illustrious career, she died of cancer in 1958, only five years after she published her work on DNA. The rules for Nobel Prizes prohibit posthumous awards, so Franklin did not receive a distinction that she richly deserved.

BARBARA McCLINTOCK DISCOVERS TRANSPOSABLE ELEMENTS.

The excitement over DNA in the 1950s overshadowed one of the most important advances in genetics: Barbara McClintock's discovery of transposable elements. Genes usually mutate at a rate of about one per million, and once a gene has mutated, the mutation nearly always remains stable, replicating itself faithfully from one generation to the next. Therefore, Barbara McClintock was dismayed when she observed frequent and unstable mutations in some of her corn plants. The mutations had been stable, but in certain crosses the mutations reverted at an unusually high rate to their original, nonmutant types. McClintock found that when she crossed two particular yellow-kernel types, the progeny kernels had dark purple spots on a yellow background. Her experimental data suggested to her that the purple spots arose when a mutated gene reverted back to its nonmutant form. Each spot on the kernel represents a mutation in one cell, and each kernel had numerous spots. Therefore, the same gene must have reverted back to the functional form in the same way in many separate cells.

Ultimately, through well-conceived experiments and a process of brilliant logic, she arrived at a stunning explanation of her results. A piece of the chromosome had inserted itself into the gene in the chromosome, disrupting the gene's function and creating a mutated gene. When the piece of chromosome excised itself, the gene's function was restored.

McClintock published a major paper on transposable

elements in 1950,[33] then presented her results at a symposium in 1951. Unfortunately, few scientists were able to follow her stunning but complicated logic. In fact, most dismissed her work as invalid, not because her interpretations were unsupported but because her conclusions challenged a long-standing view of the constancy of genes.

In retrospect, today's geneticists have little trouble understanding McClintock's articles, largely because the mechanism of transposition at the DNA level is now known. When her arguments are viewed in the context of DNA, they make perfect sense. However, she arrived at her conclusions without the assistance of any information about DNA. As it turned out, later studies with DNA would prove her conclusions as correct to the smallest detail, confirming that she was a genius in the truest sense.

In 1953, Watson and Crick unintentionally eclipsed McClintock's work as their discovery of the structure of DNA took center stage. However, McClintock's conclusions did not pass entirely unnoticed. Some of her closest colleagues knew that she was onto something big, and they encouraged her to press forward in the face of neglect and ignorance from much of the scientific community. Nearly three decades elapsed before her work was widely recognized. As it became apparent that transposable elements are probably present in all species, hundreds of geneticists rushed to study them. In 1983, thirty-three years after her landmark paper, McClintock received a Nobel Prize for her discovery of transposable elements, a long-overdue recognition of her genius.

"DNA → RNA → PROTEIN" IS THE CENTRAL DOGMA OF MOLECULAR GENETICS.

In the mid-1950s, nearly everyone had accepted the idea that DNA is the genetic material. Because proteins are the products of genes and genes are made of DNA, the linear sequence of nucleotides in the DNA of a gene must somehow encode the linear sequence of amino acids in a protein. Presumably a protein could be synthesized directly on a DNA molecule, and some biologists offered models of how this could happen. However, as information accumulated, it became clear that there must be some sort of intermediate between DNA and protein.

Elliot Volkin and Lazarus Astrachan, in several studies published between 1956 and 1958, showed that when viruses infect bacterial cells, a new type of RNA appears, and the composition of that RNA resembles the composition of the virus DNA.[34] In 1961, Sol Spiegelman found that the RNA in bacterial cells infected by viruses formed base pairs along their full lengths with virus DNA but not with bacterial DNA.[35] Therefore, RNA was a prime candidate as the intermediate. It was easy for scientists to envision how a single strand of RNA could be made from a DNA template: an enzyme could bind to the DNA and synthesize a strand of RNA using the base-pairing rules.

In 1961, François Jacob and Jacques Monod of the Pasteur Institute proposed the hypothesis of *messenger RNA (mRNA)*.[36] Sydney Brenner of Cambridge University in England, François Jacob of the Pasteur Institute in Paris, and Matthew Meselson of the California Institute of Technology coauthored a paper, which has since become a classic.[37] By radioactively labeling virus RNA, they discovered the existence of mRNA, a single-stranded RNA copy of a linear gene in DNA that carries information from DNA to make proteins.

These observations were powerful evidence of mRNA as the intermediate between DNA and protein. The direct product of a gene must be RNA, and protein must be the product encoded by the RNA. Francis Crick named this idea *the central dogma* of molecular biology:

$$DNA \rightarrow RNA \rightarrow protein$$

The arrows signify the transfer of information from one molecule to another. The first step, DNA → RNA, was named *transcription* because it represents an enzyme copying information from one molecule to another in the same "language" of bases. The second step, RNA → protein, was named *translation* because it denotes a transfer of information from one "language" to another, from bases to amino acids. The proteins encoded by DNA and RNA then direct the formation and function of an entire organism. Expressing this concept, Jeff Wheelwright, a reporter for *Life*, took the central dogma one step further: "DNA, RNA, and protein—this is the holy trinity of molecular biology. The relationship between these three molecules underlies the whole science. . . . DNA makes RNA, RNA makes protein, and protein makes us."[38]

EXPERIMENTAL MOLECULAR BIOLOGISTS CRACK THE GENETIC CODE.

There are four bases in DNA and RNA and twenty amino acids in proteins. Also, the linear sequence of bases in DNA corresponds directly to the linear sequences of bases in RNA. Does the linear sequence of bases in RNA then correspond to the linear sequence of amino acids in a protein, and if so, how? The relationship cannot be one-to-one because if it were, only four amino acids could be encoded. A two-to-one relationship also doesn't work because there are sixteen possible combinations of two bases, leaving four amino acids unspecified. The number of bases per amino acid must be at least three.

By the mid-1960s, all accumulated evidence pointed to triplet units of bases, called *codons*, in mRNA that specify amino acids. However, the *genetic code*, the relationship between specific codons and their amino acids, remained a mystery. Cracking the genetic code became the top priority for experimental molecular biologists. Several researchers working in different laboratories collaborated in an effort that revealed the entire code in the five-year period from 1961 to 1966. Marshall Nirenberg, Heinrich Matthaei, Severo Ochoa, Philip Leder, and Har Gobind Khorana led research groups that through a long series of complicated but elegant experiments determined the genetic code, which turned out to be nearly universal throughout life.[39] The universality of the genetic code suggests that it evolved very early during life's history. This discovery also has a practical application that turned out to be a tremendous benefit for medicine and for research. A gene from one organism transferred to another produces the same protein. Today, important proteins, such as human insulin used to treat diabetes, can be produced commercially in bacteria because of the universality of the genetic code.

EARLY STUDIES OF MOLECULAR EVOLUTION FOCUS ON PROTEINS.

In the late 1960s, DNA-sequencing methods were in their infancy. They were exceptionally cumbersome, and only small segments of DNA could be reliably sequenced. Protein sequencing, the determination of the amino acid sequence of a protein, was also cumbersome but at the time more advanced than DNA sequencing. Cytochrome *c* is a relatively small protein (104 amino acids in the human version) that is present in a wide array of species, including animals, plants, and microorganisms. By the late 1960s, the complete amino acid sequence of cytochrome *c* had been determined in more than twenty species.

These sequences opened the door to construction of phylogenetic trees based purely on amino acid sequences. In 1967, Walter Fitch and Emanuel Margoliash published a phylogenetic tree based entirely on cytochrome *c* sequences from mammals, birds, reptiles, fish, insects, and fungi.[40] The tree portrayed evolutionary relationships that were essentially the same as those determined by anatomical comparisons, adding molecular evidence to the reconstruction of evolutionary histories.

RELIABLE DNA-SEQUENCING METHODS ARE DEVELOPED.

By 1975, a good method for DNA sequencing was still elusive. Frederick Sanger, at Cambridge in England, had developed methods for sequencing RNA, and scientists found that they could indirectly sequence a DNA fragment by transcribing it first into an RNA molecule, then sequencing the RNA. In 1975, Sanger published a method for direct DNA sequencing called the plus-minus method that relies on gradual degradation of the DNA molecule during the sequencing process. He used it in 1977 to determine the entire DNA sequence (5,386 base pairs) of the bacterial virus φX174.[41] This was a monumental task at the time, doable now in less than a day.

In 1977, Allan Maxam and Walter Gilbert, at Harvard University, developed a method for DNA sequencing that also relies on degradation of the DNA molecule being sequenced.[42] Shortly afterward, Sanger developed yet another DNA sequencing method, called the dideoxy or chain-termination method.[43] Sanger dideoxy sequencing soon became the method of choice. Although it was expensive and labor-intensive at the time, it was less so than any other method available. Its importance grew dramatically during the 1980s when it became the basis for rapid and inexpensive automated DNA sequencing.

THE FIRST HUMAN GENES ARE SEQUENCED.

A major milestone in genetics was reached when the first human genes were sequenced. Among the most important and informative were the human insulin (*INS*) gene and the human beta-globin (*HBB*) gene.[44] The *INS* and *HBB* genes were among the first human genes sequenced because they are some of the smallest human genes. Each has only two introns and encodes a relatively small number of amino acids (110 encoded by the *INS* gene and 147 encoded by the *HBB* gene). Most human genes are much larger, with tens of thousands of nucleotides, numerous introns, and hundreds of codons in the reading frames.

INTRONS INTERRUPT EUKARYOTIC GENES.

Bacterial genes sequenced in the mid-1970s revealed that the DNA of these genes corresponds exactly with the mRNA sequence with no intervening sequences. Therefore, molecular geneticists were at first incredulous, then dismayed, when researchers from several laboratories reported in 1977–78 that intervening sequences interrupt the DNA of eukaryotic genes. These intervening sequences, which at first appeared to be useless DNA, were named *introns*.

Immediately, scientists began to wonder why introns exist and were at a loss to find any reasonable explanation for their presence. Some scientists labeled introns as "junk DNA" because introns seemed to perform no function whatsoever. Eventually, several roles for introns emerged. A few introns contain DNA sequences that help regulate their gene's function. In other cases, a single gene can encode more than one protein through alternative intron removal. Introns also appear to have played an important role in the long-term evolution of genes.

THE AGE OF GENOMICS IS BORN.

As DNA-sequencing methods were refined, the opportunities for researchers to compare sequences among species increased dramatically. Much of the early evidence on transposable elements and pseudogenes documented in chapters 2 and 3 came from DNA-sequencing studies conducted in the late 1970s and early '80s. In 1985, Robert Shinsheimer, chancellor of the University of California at Santa Cruz, convened a meeting to plan for a human genome sequencing center that he hoped to establish at his university. Although the center never materialized there, the idea was on the table, and scientists were seriously discussing the possibility of sequencing the human genome. On one thing they were all agreed—DNA-sequencing methods, and computerized storage, retrieval, and analysis systems, needed considerable improvement and automation for genome sequencing to be possible. Subsequent meetings were held the following year with much discussion and debate. David Baltimore, a Nobel laureate and one of the major players, summed up the situation, "The idea is gathering momentum. I shiver at the thought."[45]

The first successful prototype of an automated DNA sequencer was developed in 1986 at the California Institute of Technology under the direction of Leroy Hood. The following year, a small start-up biotech company that was destined to expand over the next few years, Automated Biosytems, Inc. (ABI), released the first commercial automated sequencer.

Recognizing the very real possibility of a human genome project, the National Research Council (NRC) established a committee of stellar biologists to evaluate the prospects and recommend how such a project should proceed. The committee began in 1986 and met periodically over a period of two years before issuing its report. The committee ended up endorsing a human genome project but also recommended that the project include genome projects of model organisms, such as the fruit fly, *Drosophila melanogaster*, that Morgan had chosen as a model organism eight decades earlier.

Congress appropriated funding for a human genome project at the National Institutes of Health (NIH) based on the NRC committee's recommendations. James Watson was appointed director of the project in 1988 and accepted the appointment with the words "I would only once have the opportunity to let my scientific life encompass a path from the double helix to the three billion steps of the human genome."[46] In the meantime, the Department of Energy (DOE) had devoted significant funding toward its own human genome project. DOE scientists had substantial experience in studying the effect of radiation on DNA and in management of large-scale scientific databases. Such expertise proved to be especially valuable to the project. Also in 1988, the two agencies joined forces. Other countries also initiated their own genome projects, the largest being the Sanger Centre in England, ultimately cooperating with the US project. The project eventually took on the name the *International Human Genome Sequencing Consortium*, a well-chosen title. The project officially began in 1990, although much work had already been completed by that time.

The project was not without its critics. Some scientists felt that the genome project was drawing funding away from other worthy projects, including their own. Others felt that the value of information obtained from the project would not justify the cost. Regarding this critical time in the project's history, author Robert Cook-Deegan wrote, "Watson spent days in phone-to-phone combat, turning his considerable energies to shim the sagging fate of his program. The same ardor that assembled the double helix from cardboard and wire models in 1953 constructed the political structures to shore up the NIH genome project in 1990."[47] Watson's efforts were successful, and the project moved forward in the face of criticism with healthy funding.

Even with automated DNA sequencing, it is still impossible for scientists to sequence the DNA from an entire chromosome end-to-end. Instead, they must sequence the DNA in relatively small fragments and assemble the fragments to identify a continuous sequence. The approach taken by the genome project was to assemble the DNA fragments first, then sequence them after they had been assembled. Using a combination of methods, including genetic mapping of the type developed by Sturtevant in 1911 but working directly with DNA, project scientists gradually began to assemble DNA fragments in preparation for large-scale sequencing.

In the meantime, progress on DNA-database management and development of high-throughput automated sequencers moved at lightning speed. In 1988, Congress established the National Center for Biotechnology Information to manage the large amounts of sequence data that had been accumulating. The databases eventually made their way onto the Internet where they are now freely available.

Frustrated by conflicting priorities and bureaucratic disputes, two major players left the NIH. One was the genome project's director, James Watson; the other was a talented NIH scientist, Craig Venter. Watson had made his mark by defending the project against its harshest critics during its tumultuous early years and launching the project on a fast-paced and well-planned trajectory. His replacement, Francis Collins, received praise from all camps. Collins's scientific record was outstanding, and as an outspoken Christian, he alleviated the concerns of some who were worried about ethical and moral issues associated with the project. Venter went on to work in private industry, forming his own company and initiating a private venture to sequence the human genome as a competing project.

Venter's first company was the Institute for Genomic Research, more commonly known by its acronym TIGR ("tiger"). TIGR's human genome effort eventually was transferred to a spin-off company named Celera. In the world of corporate mergers, ABI, the leading producer of automated sequencers, merged with Perkin-Elmer, a major biotechnology company, which then merged with Celera to become PE Celera (hereafter referred to simply as Celera). These mergers gave Celera access to the latest automated sequencers. By the time the Celera genome project was fully up and running, its DNA-sequencing capacity exceeded that of the rest of the world combined.

Armed with such high-sequencing capacity, Celera determined to sequence first, then assemble the sequenced fragments later, an approach called *shotgun sequencing*, the reverse of the international consortium's approach of assembling first and then sequencing. The DNA fragments Celera sequenced were chosen mostly at random with the idea of sequencing enough fragments to ensure that practically all of the genome would be contained somewhere in the sequences, most parts of the genome being represented several times over. Computers with massive analysis capability would then identify sequenced fragments with overlapping sequences as a way of assembling the genome.

With two genome projects, one public and the other private, a race was under way, and the stakes were high. The international consortium was working to release all data freely to the public, with no restrictions on its use. In other words, the genome belonged to humanity. Celera, on the other hand, hoped to reap a profit through commercialization of the sequence. And with its tremendous sequencing capacity, Celera had a good chance of winning the race even though it was a latecomer. However, both Collins and Venter played down the notion of a race. According to Collins, "The only race involved here is the human race."[48] Venter called it "a race to impact people's lives."[49] Under these noble façades were, however, a sense of both competition and mistrust.

Celera had an unusual advantage—and an unfair one in many people's minds. Under the policy of free availability of information, the international consortium had committed to release all assembled sequence information within twenty-four hours. Thus, thousands of DNA sequences already mapped to their respective positions on human chromosomes were freely available on the Internet, courtesy of the international consortium. Celera scientists then used these sequences as anchors to help them assemble their sequence, which they then kept under proprietary secrecy. The benefit was entirely a one-way street with international consortium's public information assisting Celera in its quest to win the race.

At the beginning of 2000, as both projects were nearing completion, President Bill Clinton recognized a political storm brewing with the race. Multiple disputes, including the conflict over free access to sequence information as opposed to private patents on it, made their way into the press. Clinton assigned his science adviser to bring Collins and Venter together. The human genome was too important for it to be completed under a shroud of controversy. Through a series of private meetings over pizza and beer, Collins and Venter arrived at agreements that resolved most, albeit not all, of the disputes.

On June 26, 2000, President Clinton, with Collins and Venter at his side, and with British prime minister Tony Blair representing the Sanger Centre via satellite, announced, "We are here to celebrate the completion of the first survey of the entire human genome. Without a doubt, this is the most important, most wondrous map ever produced by humankind."[50] A century almost to the day had passed since the rediscovery of Mendel's laws.

We have traversed a long and productive road since Darwin, Mendel, and Miescher. The genomic era is now in full force, with dozens of genome projects at or near completion. They continue to reveal in astonishing detail the secrets of life, including remarkable insights into our own origins and our shared genetic heritage with the rest of life.

NOTES

1. From a letter by Charles Darwin to Charles Lyell, June 18, 1858. "The Darwin Correspondence Online Database," http://darwin.lib.cam .ac.uk/perl/nav?pclass=letter&pkey=2285 (accessed January 22, 2007).

2. V. Orel, *Gregor Mendel: The First Geneticist* (Oxford: Oxford University Press, 1996), p. 71.

3. C. Stern and E. R. Sherwood, *The Origin of Genetics: A Mendel Source Book* (San Francisco: W. H. Freeman and Co., 1966), p. 2.

4. R. Olby, *Origins of Mendelism*, 2nd ed. (Chicago: University of Chicago Press, 1985), pp. 202–203.

5. Orel, *Gregor Mendel*, p. 276.

6. D. J. Fairbanks and B. Rytting, "Mendelian Controversies: A Botanical and Historical Review," *American Journal of Botany* 88 (2001): 737–52.

7. W. O. Focke, *Die Pflanzen-Mischling. Ein Beitrag zur Biologie der Gewächse* (Berlin: Gebruder Borträger, 1881).

8. Fairbanks and Rytting, "Mendelian Controversies," p. 750.

9. F. Darwin, ed., *The Autobiography of Charles Darwin and Selected Letters* (New York: Dover Publications, 1958), p. 49.

10. Orel, *Gregor Mendel*, p. 194

11. Emma Darwin to Henrietta Darwin, March 19, 1870, as quoted in K. Pearson, *Life, Letters, and Labours of Francis Galton*, vol. 2 (Cambridge: Cambridge University Press, 1924), p. 158.

12. B. E. Bishop, "Mendel's Opposition to Evolution and Darwin," *Journal of Heredity* 87 (1996): 205–13; L. A. Callender, "Gregor Mendel: An Opponent of Descent with Modification," *History of Science* 26 (1988): 41–57.

13. Orel, *Gregor Mendel*; Fairbanks and Rytting, "Mendelian Controversies."

14. E. B. Wilson, *The Cell in Development and Heredity*, 2nd ed. (London: Macmillan, 1900), p. 358.

15. Some historians have questioned whether or not Mendel's work was truly rediscovered. For Correns's, de Vries's and Tschermak's work to correctly be called a rediscovery, they must have arrived at the same conclusions from their experiments as Mendel

did, but before seeing his work. Some historians claim that they did not recognize Mendelian laws until after reading his article. However, each of them recalled their interpretation as predating their reading of Mendel's paper.

16. A. H. Sturtevant, *A History of Genetics* (New York: Harper and Row Publishers, 1965), p. 47.

17. L. V. Morgan, "Non Criss-Cross Inheritance in *Drosophila Melanogaster*," *Biological Bulletin* 42 (1922): 267–74; C. B. Bridges, "Sex in Relation to Chromosomes and Genes," *American Naturalist* 59 (1925): 127–37.

18. H. B. Creighton and B. McClintock, "A Correlation of Cytological and Genetical Crossing Over in *Zea Mays*," *Proceedings of the National Academy of Sciences, USA* 17 (1931): 492–97.

19. C. B. Bridges, "The Translocation of a Section of Chromosome-II upon Chromosome-III in *Drosophila*," *Anatomical Record* 24 (1923): 426–27; Sturtevant, "A Crossover Reducer in *Drosophila Melanogaster* Due to Inversion of a Section of the Third Chromosome," *Biologisches Zentralblatt* 46 (1926): 697–702.

20. J. H. Bennett, from the foreword of the 1999 edition of R. A. Fisher, *The Genetical Theory of Natural Selection* (Oxford: Oxford University Press, 1930), p. vi.

21. Leonard Darwin to Ronald A. Fisher, June 9, 1930, in J. H. Bennett, ed., *Natural Selection, Heredity, and Eugenics* (Oxford: Clarendon Press, 1983), p. 121.

22. S. Wright, "Evolution in Mendelian Populations," *Genetics* 16 (1931): 97–159.

23. S. Wright, *Evolution and the Genetics of Populations*, vols. 1–4 (Chicago: University of Chicago Press, 1968–78).

24. A. D. Hershey and M. Chase, "Independent Functions of Viral Protein and Nucleic Acid in Growth of Bacteriophage," *Journal of General Physiology* 36 (1952): 39–56.

25. E. Chargaff, "Structure and Function of Nucleic Acids as Cell Constituents," *Federation Proceedings* 10 (1951): 659.

26. F. H. Portugal and J. S. Cohen, *A Century of DNA: A History of the Structure and Function of the Genetic Substance* (Cambridge, MA: MIT Press, 1977), p. 249.

27. J. D. Watson, *The Double Helix: A Personal Account of the Discovery of the Structure of DNA* (New York: W. W. Norton & Co., 1980), p. 60.

28. Ibid., p. 108.

29. R. Olby, "Francis Crick, DNA, and the Central Dogma," *Daedalus* 99 (1970): 958.

30. R. Olby, *The Path to the Double Helix* (Seattle: University of Washington Press, 1974), p. 417.

31. J. D. Watson and F. H. C. Crick, "A Structure for Deoxyribose Nucleic Acid," *Nature* 171 (1953): 737–38.

32. R. Franklin and R. G. Gosling, "Molecular Configuration in Sodium Thymonucleate," *Nature* 171 (1953): 740–41; M. H. F. Wilkins, "Molecular Structure of Desoxypentose Nucleic Acids," *Nature* 171 (1953): 738–40.

33. B. McClintock, "The Origin and Behavior of Mutable Loci in Maize," *Proceedings of the National Academy of Sciences, USA* 36 (1950): 344–55.

34. E. Volkin and L. Astrachan, "Intracellular Distribution of Labeled Ribonucleic Acid after Phage Infection of *Escherichia coli*," *Virology* 2 (1956): 433–37; ibid., "Phosphorus Incorporation in *Escherichia coli* Ribonucleic Acid after Infection with Bacteriophage T2," *Virology* 2 (1956): 149–61; ibid., "Properties of Ribonucleic Acid Turnover in T2-Infected *Escherichia coli*," *Biochimica et Biophysica Acta* 29 (1958): 536–44.

35. S. Spiegelman, B. D. Hall, and R. Storck, "The Occurrence of Natural DNA-RNA Complexes in *E. coli* Infected with T2," *Proceedings of the National Academy of Sciences, USA* 47 (1961): 1135–41.

36. F. Jacob and J. Monod, "Genetic Regulatory Mechanisms in the Synthesis of Proteins," *Journal of Molecular Biology* 3 (1961): 318–56.

37. S. Brenner, F. Jacob, and M. Meselson, "An Unstable Intermediate Carrying Information from Genes to Ribosomes for Protein Synthesis," *Nature* 190 (1961): 576–81.

38. J. Wheelwright, "Weaving the New Threads of Life," *Life* 3, no. 5 (1980): 52.

39. C. R. Woese, *The Genetic Code: The Molecular Basis for Genetic Expression* (New York: Harper & Row Publishers, 1967).

40. W. M. Fitch and E. Margoliash, "Construction of Phylogenetic Trees," *Science* 155 (1967): 279–84.

41. F. Sanger et al., "Nucleotide Sequence of Bacteriophage Phi X174 DNA," *Nature* 265 (1977): 687–95.

42. A. M. Maxam and W. Gilbert, "A New Method for Sequencing DNA," *Proceedings of the National Academy of Sciences, USA* 74 (1977): 560–64.

43. F. Sanger, S. Nicklen, and A. R. Coulson, "DNA Sequencing with Chain-Terminating Inhibitors," *Proceedings of the National Academy of Sciences, USA* 74 (1977): 5463–67.

44. G. I. Bell et al., "Sequence of the Human Insulin Gene," *Nature* 284 (1980): 26–32; R. M. Lawn et al., "The Nucleotide Sequence of the Human Beta-Globin Gene," *Cell* 21 (1980): 647–51.

45. R. Lewin, "Proposal to Sequence the Human Genome Stirs Debate," *Science* 232 (1986): 1600.

46. R. Cook-Deegan, *The Gene Wars: Science, Politics, and the Human Genome* (New York: W. W. Norton & Co., 1994), p. 161.

47. Ibid., p. 175.

48. F. Golden and M. D. Lemonick, "The Race Is Over," *Time* 156, no. 1 (2000), p. 23.

49. Ibid.

50. White House, Office of the Press Secretary, "Text of Remarks on the Completion of the First Survey of the Entire Human Genome Project," http://clinton5.nara.gov/WH/New/html/genome-20000626.html (accessed January 27, 2007).

GLOSSARY

alphoid sequence: A repeated sequence of DNA found in centromeres.

***Alu* element:** A relatively small retroelement that is the most common transposable element in the genomes of primates.

apes: A subgroup of primates that includes chimpanzees, gorillas, orangutans, and gibbons.

artificial selection: The intentional action by humans with animals and plants of choosing which parents to breed with the intent of improving inherited characteristics.

base: The chemical component of a DNA or RNA molecule that varies. There are four bases in DNA: thymine (T), cytosine (C), adenine (A), and guanine (G), and four bases in RNA: uracil (U), cytosine (C), adenine (A), and guanine (G).

center of origin: The geographic region where a particular species originated.

central dogma: The information transfer from DNA to RNA to protein through transcription and translation.

centromere: A microscopically visible constricted region in a chromosome that contains alphoid sequences and serves as a site for the cell to direct chromosomes to their proper positions during cell division.

chloroplast: A structure found in plant cells that carries out photosynthesis and contains its own DNA. Chloroplasts resemble photosynthetic bacteria and, according to the endosymbiotic hypothesis, trace their evolutionary origin to them.

chromosome: A microscopic structure found in the cell nucleus (or simply in the cell, in the case of bacteria) that consists of a DNA molecule stabilized by proteins.

chromosome rearrangement: A major alteration in a chromosome caused by a rearrangement of part of the chromosome's structure. Rearrangements include inversions, translocations, fusions, fissions, deletions, and duplications of chromosomal material.

coevolution: The evolution of one species as affected by the evolution of another. Coevolution in two species is usually mutual.

creationism: A religious and political movement whose adherents claim that scientific evidence supports a literal interpretation of the biblical account of creation.

Darwinism: The idea that the major mechanism of evolution is natural selection.

deletion: Elimination of part of a DNA molecule, ranging from a single base pair to a large segment of a chromosome containing as many as millions of base pairs.

DNA: Deoxyribonucleic acid, a molecule that carries the information of heredity from one cell generation to the next and one organismal generation to the next.

duplication: Copying of a segment of DNA followed by insertion of that segment into the DNA molecule, leaving the original segment intact. Many duplications are present as tandem repeats of DNA.

duplication pseudogene: A mutated, nonfunctional gene that arose by duplication of a gene followed by mutation.

endosymbiotic hypothesis: The hypothesis, supported by

abundant evidence, that mitochondria and chloroplasts trace their ancestry back more than a billion years to free-living bacterial cells.

eukaryotes: Organisms whose cells contain a nucleus.

evolution: Genetic change in populations eventually resulting in the divergence of genetic lineages of the same species into separate species.

exon: A segment of DNA in a gene that remains in the final version of the RNA encoded by that gene.

fission: Breakage of a single chromosome to form two chromosomes.

fusion: Merger of two chromosomes to form a single chromosome.

gene: A segment of DNA that encodes an RNA, which usually encodes a protein.

gene conversion: The transfer of a segment of one gene into a second gene by virtue of highly similar DNA sequences in the two genes.

Genographic Project: A project sponsored by the National Geographic Society aimed at increasing the DNA database of genetic variations among indigenous peoples of the world with the objective of clarifying ancient human migrations.

genome: The complete set of genetic information that an organism inherits from one parent. One genome contains one set of chromosomes, and each egg or sperm cell typically carries one genome. Human cells, except for egg and sperm cells, carry two genomes, one inherited from the mother and the other from the father.

germline: The cells in embryos and reproductive organs that give rise to egg or sperm cells.

great apes: The larger apes, including chimpanzees, gorillas, and orangutans but excluding gibbons.

hemoglobin: An iron-containing, red-colored blood protein that carries oxygen from the lungs to tissues in the body and carbon dioxide from the tissues to the lungs.

Hominidae: A classification that traditionally included humans and extinct humanlike species but excluded the great apes. The classification has been revised on the basis of DNA evidence to include humans and great apes (chimpanzees, gorillas, and orangutans).

Human Genome Project: A project that determined the DNA sequence of the human genome and annotated it to identify the positions of genes and other features, such as transposable elements and pseudogenes.

inheritance of acquired characteristics: A discredited notion that natural and artificial selection directly modify the genetic material to change the characteristics of inherited traits.

in-silico research: Research conducted on DNA, RNA, and protein sequences contained in large, computerized databases.

intelligent design: A political and quasi-religious movement claiming that the complexity of living organisms can be explained only by the existence of an intelligent agent who designed them.

interspecific hybrid: Offspring of a mating between two different but related species. Many interspecific hybrids are infertile, mules being the best-known example.

intron: A segment of DNA in a gene that is copied into RNA but removed from the RNA before the RNA transmits its information to proteins.

inversion: A chromosomal segment that has been inverted relative to its original conformation.

irreducible complexity: Complexity of a biological system that cannot function when one of its multiple parts fails to function. Intelligent design advocates claim that irreducible complexity is evidence of design, whereas most scientists reject the notion that irreducible complexity can only be explained by intelligent design.

mitochondrion (plural *mitochondria*): A structure residing outside of the nucleus in eukaryotic cells that carries its own DNA and resembles bacterial cells. According to the endosymbiotic hypothesis, mitochondria trace their ancestral origin to free-living bacteria that entered primitive eukaryotic cells more than a billion years ago.

Modern Synthesis: The synthesis during the 1930s of the Darwinian theory of evolution by natural selection with the Mendelian theory of inheritance.

mutation: An inherited change in DNA sequence. Types of mutations include substitutions, insertions, or deletions of base pairs in DNA, as well as insertions of transposable elements and pseudogenes, and chromosomal rearrangements.

natural selection: The process by which organisms that are best adapted to their surroundings are most likely to survive and reproduce, passing on the inherited traits that confer the ability to survive and reproduce to their offspring.

neo-Darwinism: The modern theory of evolution by natural selection as informed by the Mendelian theory of inheritance.

nucleus: A membrane-bound compartment in eukaryotic cells that contains the chromosomes.

Pongidae: The now-defunct classification of the great apes as a single genetic group distinct from humans. On the basis of DNA evidence, biologists now include the great apes and humans in a single group called the Hominidae.

primates: A scientific group of related species that includes humans, apes, monkeys, and lemurs.

prokaryotes: Species that lack a cell nucleus. Includes all of the bacteria.

protein: A linear chain or chains of amino acids folded into a three-dimensional structure that carries out a biological function.

pseudogene: A copy of a gene or portion of a gene that does not function.

recombination: Exchange of DNA segments between two DNA molecules that are highly similar in sequence.

replication: Copying of one double-stranded DNA molecule to make two identical double-stranded molecules.

retroelement: A transposable element that is copied from DNA into RNA as an intermediate and then back again into DNA.

retropseudogene: A pseudogene that arose when the RNA molecule transcribed from a gene is reverse transcribed back into DNA and the DNA inserted into the genome.

retrovirus: A virus composed of RNA that is reverse transcribed into DNA, which can be transcribed to make more RNA copies of the virus.

RNA: Ribonucleic acid, a molecule typically copied from a gene in DNA that is very similar to DNA in composition.

selection: Favoring of particular individuals over others for mating, either through intentional human choice (artificial selection) or by ability to survive and reproduce in nature (natural selection).

selectively neutral mutation: A mutation that is neither favored nor disfavored by selection.

selectively relevant mutation: A mutation that is either favored or disfavored by selection.

special creation: The doctrine that each species was separately created by God and has remained unchanged since the time of creation.

species: A group of genetically similar individuals that can mate and produce fertile offspring.

telomere: Segment of DNA at each end of a chromosome that consists of a series of six base-pair tandem repeats.

transcription: Copying of a DNA molecule into single-stranded RNA.

translation: Assembly of a chain of amino acids from an RNA template according to the genetic code to form a protein.

translocation: Transfer of a chromosome segment from one chromosome to a different chromosome. Most translocations are reciprocal, an exchange of segments between different chromosomes.

transposable element: A segment of DNA that can either be excised and moved to another location (transposon) or copied into an RNA intermediate and then copied back into DNA, which is then inserted into the genome (retroelement).

transposon: A transposable element composed of DNA that can be excised and moved to another location in the genome.

Tree of Life Project: A project sponsored by the National Science Foundation to determine the evolutionary relationships of many species on the basis of large-scale analysis of DNA sequence information.

trichotomy problem: The problem of identifying among humans, chimpanzees, and gorillas which two are most closely related to each other. DNA evidence has solved the trichotomy problem, clearly showing that humans and chimpanzees are more closely related to each other than either is to gorillas.

unitary pseudogene: A pseudogene that arose when a gene mutated into a nonfunctional form, and no functional copy of that gene is present in the species.

variety: A genetically distinct group of individuals in a species that is usually isolated to some degree from other such groups but that can still mate with other groups and produce fertile offspring. Varieties are the precursors of new species.

virus: A molecule of DNA or RNA encapsulated in a protein coat that can enter a cell and replicate to make multiple copies of itself.

BIBLIOGRAPHY

Arnason, U., A. Gullberg, S. Gretarsdottir, B. Ursing, and A. Janke. "The Mitochondrial Genome of the Sperm Whale and a New Molecular Reference for Estimating Eutherian Divergence Dates." *Journal of Molecular Evolution* 50 (2000): 569–78.

Astrachan, L., and E. Volkin. "Properties of Ribonucleic Acid Turnover in T2-Infected *Escherichia coli.*" *Biochimica et Biophysica Acta* 29 (1958): 536–44.

Avarello, R., A. Pedicini, A. Caiuolo, O. Zuffardi, and M. Fraccaro. "Evidence for an Ancestral Alphoid Domain on the Long Arm of Human Chromosome 2." *Human Genetics* 89 (1992): 247–49.

Baldini, A., T. Ried, V. Shridhar, K. Ogura, L. D'Aiuto, M. Rocchi, and D. C. Ward. "An Alphoid Sequence Conserved in All Human and Great Ape Chromosomes: Evidence for Ancient Centromeric Sequences at Human Chromosomal Regions 2q21 and 9q13." *Human Genetics* 90 (1993): 577–83.

Behe, M. J. *Darwin's Black Box: The Biochemical Challenge to Evolution.* New York: Simon & Schuster, 1996.

Bell, G. I., R. L. Pictet, W. J. Rutter, B. Cordell, E. Tischer, and H. M. Goodman. "Sequence of the Human Insulin Gene." *Nature* 284 (1980): 26–32.

Bennett, J. H., ed. *Natural Selection, Heredity, and Eugenics.* Oxford: Clarendon Press, 1983.

Bishop, B. E. "Mendel's Opposition to Evolution and Darwin." *Journal of Heredity* 87 (1996): 205–13.

Booth, H. A. F., and P. W. H. Holland. "Eleven Daughters of *NANOG*." *Genomics* 84 (2004): 229–38.

Brenner, S., F. Jacob, and M. Meselson. "An Unstable Intermediate Carrying Information from Genes to Ribosomes for Protein Synthesis." *Nature* 190 (1961): 576–81.

Bridges, C. B. "Sex in Relation to Chromosomes and Genes." *American Naturalist* 59 (1925): 127–37.

———. "The Translocation of a Section of Chromosome-II upon Chromosome-III in *Drosophila*." *Anatomical Record* 24 (1923): 426–27.

Britten, R. J. "Coding Sequences of Functioning Human Genes Derived Entirely from Mobile Element Sequences." *Proceedings of the National Academy of Sciences, USA* 101 (2004): 16825–30.

Callender, L. A. "Gregor Mendel: An Opponent of Descent with Modification." *History of Science* 26 (1988): 41–57.

Cann, R. L., M. Stoneking, and A. C. Wilson. "Mitochondrial DNA and Human Evolution." *Nature* 325 (1987): 31–36.

Cavalli-Sforza, L. L., and F. Cavalli-Sforza. *The Great Human Diasporas: The History of Diversity and Evolution*. Reading, MA: Addison-Wesley Publishing Co. Inc., 1995.

Chang, L.-Y. E., and J. L. Slightom. "Isolation and Nucleotide Sequence Analysis of the β-type Globin Pseudogene from Human, Gorilla, and Chimpanzee." *Journal of Molecular Biology* 180 (1984): 767–84.

Chargaff, E. "Structure and Function of Nucleic Acids as Cell Constituents." *Federation Proceedings* 10 (1951): 654–59.

Charlesworth, B. "The Organization and Evolution of the Human Y Chromosome." *Genomic Biology* 4 (2003): 226.

Charlesworth, D., B. Charlesworth, and G. Marais. "Steps in the Evolution of Heteromorphic Sex Chromosomes." *Heredity* 95 (2005): 118–28.

Chen, F.-C., and W.-H. Li. "Genomic Divergence between Humans and Other Hominoids and the Effective Population Size of the Common Ancestor of Humans and Chimpanzees." *American Journal of Human Genetics* 68 (2001): 444–56.

Chen, Y. S., A. Torroni, L. Excoffier, A. S. Santachiara-Benerecetti, and D. C. Wallace. "Analysis of mtDNA Variation in African Populations Reveals the Most Ancient of All Human Continent-Specific Haplogroups." *American Journal of Human Genetics* 57 (1995): 133–49.

Chimpanzee Sequencing and Analysis Consortium. "Initial Sequence of the Chimpanzee Genome and Comparison with the Human Genome." *Nature* 437 (2005): 69–87.

Collins, F. S. *The Language of God: A Scientist Presents Evidence for Belief.* New York: Free Press, 2006.

Collodel, G., E. Moretti, S. Capitani, P. Piomboni, C. Anichini, M. Estenoz, and B. Baccetti. "TEM, FISH and Molecular Studies in Infertile Men with Pericentric Inversion of Chromosome 9." *Andrologia* 38 (2006): 122–27.

Comas, D., S. Plaza, R. S. Wells, N. Yuldaseva, O. Lao, F. Calafell, J. Bertranpetit. "Admixture, Migrations, and Dispersals in Central Asia: Evidence from Maternal DNA Lineages." *European Journal of Human Genetics* 12 (2004): 495–504.

Cook-Deegan, R. *The Gene Wars: Science, Politics, and the Human Genome.* New York: W. W. Norton & Co., 1994.

Creighton, H. B., and B. McClintock. "A Correlation of Cytological and Genetical Crossing Over in *Zea Mays.*" *Proceedings of the National Academy of Sciences, USA* 17 (1931): 492–97.

"Darwin Correspondence Online Database." http://darwin.lib.cam.ac .uk/perl/nav?pclass=letter&pkey=2285 (accessed January 22, 2007).

Darwin, C. E. *The Descent of Man, and Selection in Relation to Sex.* 2nd ed. London: John Murray, 1882.

———. *On the Origin of Species by Means of Natural Selection or the Preservation of Favoured Races in the Struggle for Life.* 6th ed. London: John Murray, 1872.

Darwin, F., ed. *The Autobiography of Charles Darwin and Selected Letters.* New York: Dover Publications, 1958.

Davalos, I. P., F. Rivas, A. L. Ramos, C. Galaviz, L. Sandoval, and H. Rivera. "Inv(9)(p24q13) in Three Sterile Brothers." *Annales of Génétique* 2, no. 43 (2000): 51–54.

Dawkins, R. *The Blind Watchmaker: Why the Evidence of Evolution*

Reveals a Universe without Design. New York: W. W. Norton & Co., 1986.

———. *River Out of Eden.* New York: Basic Books, 1995.

———. *Unweaving the Rainbow: Science, Delusion and the Appetite for Wonder.* Boston: Houghton Mifflin Co., 1998.

de Tocqueville, A. *De la Démocratie en Amérique, I, Deuxième Partie.* http://classiques.uqac.ca/classiques/De_tocqueville_alexis/democratie_1/democratie_t1_2.pdf (accessed December 25, 2006).

Deininger, P. L., and M. A. Batzer. "*Alu* Repeats and Human Disease." *Molecular Genetics and Metabolism* 67 (1999): 183–93.

Denton, M. *Evolution: A Theory in Crisis.* Bethesda, MD: Adler and Adler, 1986.

Dobzhansky, T. "Nothing in Biology Makes Sense Except in the Light of Evolution." *American Biology Teacher* 35 (1973): 125–29.

Eikenboom, J. C., T. Vink, E. Briet, J. J. Sixma, and P. H. Reitsma. "Multiple Substitutions in the von Willebrand Factor Gene that Mimic the Pseudogene Sequence." *Proceedings of the National Academy of Sciences, USA* 91 (1994): 2221–24.

Fairbanks, D. J., and B. Rytting. "Mendelian Controversies: A Botanical and Historical Review." *American Journal of Botany* 88 (2001): 737–52.

Fairbanks, D. J., and P. J. Maughan. "Evolution of the *NANOG* Pseudogene Family in the Human and Chimpanzee Genomes." *BMC Evolutionary Biology* 6 (2006): 12. Online at http://www.biomedcentral.com/1471-2148/6/12.

Fisher, R. A. *The Genetical Theory of Natural Selection.* Oxford: Oxford University Press, 1930.

Fitch, W. M., and E. Margoliash. "Construction of Phylogenetic Trees." *Science* 155 (1967): 279–84.

Focke, W. O. *Die Pflanzen-Mischling. Ein Beitrag zur Biologie der Gewächse.* Berlin, Germany: Gebruder Borträger, 1881.

Forrest, C., and P. R. Gross. *Creationism's Trojan Horse: The Wedge of Intelligent Design.* Oxford: Oxford University Press.

Forster, P. "Ice Ages and the Mitochondrial DNA Chronology of Human Dispersals: A Review." *Philosophical Transactions of the Royal Society of London. Series B, Biological Sciences* 359 (2004): 255–64.

Franklin, R., and R. G. Gosling. "Molecular Configuration in Sodium Thymonucleate." *Nature* 171 (1953): 740–41.

Gatesy, J., C. Hayashi, M. A. Cronin, and P. Arctander. "Evidence from Milk Casein Genes that Cetaceans Are Close Relatives of Hippopotamid Artiodactyls." *Molecular Biology and Evolution* 13 (1996): 954–63.

Gibbons, R., L. J. Dugaiczyk, T. Girke, B. Duistermars, R. Zielinski, and A. Dugaiczyk. "Distinguishing Humans from Great Apes with *Alu*Yb8 Repeats." *Journal of Molecular Biology* 339 (2004): 721–29.

Godfrey, L. R., ed. *Scientists Confront Creationism.* New York: W.W. Norton & Co., 1983.

Goidts, V., J. M. Szamalek, H. Hameister, and H. Kehrer-Sawatzki. "Segmental Duplication Associated with the Human-Specific Inversion of Chromosome 18: A Further Example of the Impact of Segmental Duplications on Karyotype and Genome Evolution in Primates." *Human Genetics* 115 (2004): 116–22.

Goidts, V., J. M. Szamalek, P. J. de Jong, D. N. Cooper, N. Chuzhanova, H. Hameister, and H. Kehrer-Sawatzki. "Independent Intrachromosomal Recombination Events Underlie the Pericentric Inversions of Chimpanzee and Gorilla Chromosomes Homologous to Human Chromosome 16." *Genome Research* 15 (2005): 1232–42.

Golden, F., and M. D. Lemonick, "The Race Is Over," *Time* 156, no. 1 (2000), pp. 19–23.

Graur, D., and D. G. Higgins. "Molecular Evidence for the Inclusion of Cetaceans within the Order Artiodactyla." *Molecular Biology and Evolution* 11 (1994): 357–64.

Gvozdev, V. A., G. L. Kogan, and L. A. Usakin. "The Y Chromosome as a Target for Acquired and Amplified Genetic Material in Evolution." *Bioessays* 27 (2005): 1256–62.

Hammer, M. F. "A Recent Common Ancestry for Human Y Chromosomes." *Nature* 378 (2000): 376–78.

Hammer, M. F., D. Garrigan, E. Wood, J. A. Wilder, Z. Mobasher, A. Bigham, J. G. Krenz, and M. W. Nachman. "Heterogeneous Patterns of Variation among Multiple Human X-linked Loci: The Possible Role of Diversity-Reducing Selection in Non-Africans." *Genetics* 167 (2004): 1841–53.

Hammer, M. F., F. Blackmer, D. Garrigan, M. W. Nachman, and J. A. Wilder. "Human Population Structure and Its Effects on Sampling Y Chromosome Sequence Variation." *Genetics* 164 (2003): 1495–1509.

Hammer, M. F., T. M. Karafet, A. J. Redd, H. Jarjanazi, S. Santachiara-Benerecetti, H. Soodyall, and S. L. Zegura. "Hierarchical Patterns of Global Human Y-chromosome Diversity." *Molecular Biology and Evolution* 18 (2001): 1189–1203.

Harris, S., P. A. Barrie, M. L. Weiss, and A. J. Jeffreys. "The Primate ψβ1 Gene: An Ancient β-globin Pseudogene." *Journal of Molecular Biology* 180 (1984): 785–801.

Hershey, A. D., and M. Chase. "Independent Functions of Viral Protein and Nucleic Acid in Growth of Bacteriophage." *Journal of General Physiology* 36 (1952): 39–56.

Horai, S., and K. Hayasaka. "Intraspecific Nucleotide Differences in the Major Noncoding Region of Human Mitochondrial DNA." *American Journal of Human Genetics* 46 (1990): 828–42.

Horai, S., K. Hayasaka, R. Kondo, K. Tsugane, and N. Takahata. "Recent African Origin of Modern Humans Revealed by Complete Sequences of Hominoid Mitochondrial DNAs." *Proceedings of the National Academy of Sciences, USA* 92 (1995): 532–36.

Ijdo, J. W., A. Baldini, D. C. Ward, S. T. Reeders, and R. A. Wells. "Origin of Human Chromosome 2: An Ancestral Telomere-Telomere Fusion." *Proceedings of the National Academy of Sciences, USA* 88 (1991): 9051–55.

Inai, Y., and M. Nishikimi. "Random Nucleotide Substitutions in Primate Nonfunctional Gene for L-gulono-gamma-lactone Oxidase, the Missing Enzyme in L-ascorbic Acid Biosynthesis." *Biochimica et Biophyisica Acta* 1472 (1999): 408–11.

Inai, Y., Y. Ohta, and M. Nishikimi. "The Whole Structure of the Human Nonfunctional L-gulono-gamma-lactone Oxidase Gene—the Gene Responsible for Scurvy—and the Evolution of Repetitive Sequences Thereon." *Journal of Nutritional Science and Vitaminology* 49 (2003): 315–19.

Ingman, M., H. Kaessmann, S. Pääbo, and U. Gyllensten. "Mitochondrial Genome Variation and the Origin of Modern Humans." *Nature* 408 (2000): 708–13.

International Human Genome Sequencing Consortium. "Finishing the Euchromatic Sequence of the Human Genome." *Nature* 431 (2004): 931–45.

Ivics, Z., P. B. Hackett, R. H. Plasterk, and Z. Izsva. "Molecular Reconstruction of Sleeping Beauty, a Tc1-like Transposon from Fish, and Its Transposition in Human Cells." *Cell* 91 (1997): 501–10.

Jacob, F., and J. Monod. "Genetic Regulatory Mechanisms in the Synthesis of Proteins." *Journal of Molecular Biology* 3 (1961): 318–56.

Johnson, P. E. *Darwin on Trial.* Washington, DC: Regnery Publishing, Inc., 1991.

———. *Defeating Darwinism by Opening Minds.* Downer's Grove, IL: InterVarsity Press, 1997.

Karafet, T., L. Xu, R. Du, W. Wang, S. Feng, R. S. Wells, A. J. Redd, S. L. Zegura, and M. F. Hammer. "Paternal Population History of East Asia: Sources, Patterns, and Microevolutionary Processes." *American Journal of Human Genetics* 69 (2001): 615–28.

Kehrer-Sawatzki, H., B. Schreiner, S. Tanzer, M. Platzer, S. Muller, and H. Hameister. "Molecular Characterization of the Pericentric Inversion that Causes Differences between Chimpanzee Chromosome 19 and Human Chromosome 17." *American Journal of Human Genetics* 71 (2002): 375–88.

Kehrer-Sawatzki, H., C. A. Sandig, V. Goidts, and H. Hameister. "Breakpoint Analysis of the Pericentric Inversion between Chimpanzee Chromosome 10 and the Homologous Chromosome 12 in Humans." *Cytogenetic and Genome Research* 108 (2005): 91–97.

Kehrer-Sawatzki, H., C. Sandig, N. Chuzhanova, V. Goidts, J. M. Szamalek, S. Tanzer, S. Muller, M. Platzer, D. N. Cooper, and H. Hameister. "Breakpoint Analysis of the Pericentric Inversion Distinguishing Human Chromosome 4 from the Homologous Chromosome in the Chimpanzee (*Pan troglodytes*)." *Human Mutation* 25 (2005): 45–55.

Kehrer-Sawatzki, H., J. M. Szamalek, S. Tanzer, M. Platzer, and H. Hameister. "Molecular Characterization of the Pericentric Inversion of Chimpanzee Chromosome 11 Homologous to Human Chromosome 9." *Genomics* 85 (2005): 542–50.

Keller, E. F. *A Feeling for the Organism: The Life and Work of Barbara McClintock.* San Francisco: W. H. Freeman, 1983.

Kitzmiller et al. v. Dover Area School District, Case 4:04-cv-02688-JEJ, Document 342 (2005).

Kleineidam, R. G., G. Pesole, H. J. Breukelman, J. J. Beintema, and R. A. Kastelein. "Inclusion of Cetaceans within the Order Artiodactyla Based on Phylogenetic Analysis of Pancreatic Ribonuclease Genes." *Journal of Molecular Evolution* 48 (1999): 360–68.

Lawn, R. M., A. Efstratiadis, C. O'Connell, and T. Maniatis. "The Nucleotide Sequence of the Human Beta-Globin Gene." *Cell* 21 (1980): 647–51.

Lebedev, Y. B., O. S. Belonovitch, N. V. Zybrova, P. P. Khil, S. G. Kurdyukov, T. V. Vinogradova, G. Hunsmann, and E. D. Sverdlov. "Differences in HERV-K LTR Insertions in Orthologous Loci of Humans and Great Apes." *Gene* 247 (2000): 265–77.

Lewin, R. "Proposal to Sequence the Human Genome Stirs Debate." *Science* 232 (1986): 1598–1600.

Locke, D. P., N. Archidiacono, D. Misceo, M. F. Cardone, S. Deschamps, B. Roe, M. Rocchi, and E. E. Eichler. "Refinement of a Chimpanzee Pericentric Inversion Breakpoint to a Segmental Duplication Cluster." *Genome Biology* 4 (2003): R50.

Lucas, J. R. "Wilberforce and Huxley: A Legendary Encounter." *Historical Journal* 22 (1979): 313–30.

Maca-Meyer, N., A. M. Gonzales, J. M. Larruga, C. Flores, and V. M. Cabrera. "Major Genomic Mitochondrial Lineages Delineate Early Human Expansions." *BMC Genetics* 2 (2001): 13. Online at http://www.biomedcentral.com/1471-2156/2/13.

Macaulay, V., C. Hill, A. Achilli, C. Rengo, D. Clarke, W. Meehan, J. Blackburn, O. Semino, R. Scozzari, F. Cruciani, A. Taha, N. K. Shaari, J. M. Raja, P. Ismail, Z. Zainuddin, W. Goodwin, D. Bulbeck, H. J. Bandelt, S. Oppenheimer, A. Torroni, and M. Richards. "Single, Rapid Coastal Settlement of Asia Revealed by Analysis of Complete Mitochondrial Genomes." *Science* 308 (2005): 1034–36.

Maniou, Z., O. C. Wallis, and M. Wallis. "Episodic Molecular Evolution of Pituitary Growth Hormone in Cetartiodactyla." *Journal of Molecular Evolution* 58 (2004): 743–53.

Martin, M. J., J. C. Rayner, P. Gagneux, J. W. Barnwell, and A. Varki. "Evolution of Human-Chimpanzee Differences in Malaria Susceptibility: Relationship to Human Genetic Loss of N-glycolyl-neuraminic Acid." *Proceedings of the National Academy of Sciences, USA* 102 (2005): 12819–24.

Maxam, A. M., and W. Gilbert. "A New Method for Sequencing DNA." *Proceedings of the National Academy of Sciences, USA* 74 (1977): 560–64.

McClintock, B. "The Origin and Behavior of Mutable Loci in Maize." *Proceedings of the National Academy of Sciences, USA* 36 (1950): 344–55.

McConkey E. H. "Orthologous Numbering of Great Ape and Human Chromosomes Is Essential for Comparative Genomics." *Cytogenetic and Genome Research* 105 (2004): 157–58.

McLean v. Arkansas Board of Education. Reprinted in *Science* 215 (1982): 934–43.

Medvedev, Z. A. *The Rise and Fall of T. D. Lysenko.* Translated by I. M. Lerner. New York: Columbia University Press, 1969.

Miller, K. W. *Finding Darwin's God: A Scientist's Search for Common Ground between God and Evolution.* New York: Cliff Street Books, 1999.

Miskey, C., Z. Izsvak, R. H. Plasterk, and Z. Ivics. "The Frog Prince: A Reconstructed Transposon from *Rana pipiens* with High Transpositional Activity in Vertebrate Cells." *Nucleic Acids Research* 31 (2003): 6873–81.

Montefalcone, G., S. Tempesta, M. Rocchi, and N. Archidiacono. "Centromere Repositioning." *Genome Research* 9 (1999): 1184–88.

Morgan, L. V. "Non Criss-cross Inheritance in *Drosophila melanogaster.*" *Biological Bulletin* 42 (1922): 267–74.

Muller, S., R. Stanyon, P. C. O'Brien, M. A. Ferguson-Smith, R. Plesker, and J. Wienberg. "Defining the Ancestral Karyotype of All Primates by Multidirectional Chromosome Painting between Tree Shrews, Lemurs and Humans." *Chromosoma* 108 (1999): 393–400.

Musurillo, H. A. *The Fathers of the Primitive Church.* New York: New American Library, Inc., 1966.

National Academy of Sciences. *Teaching about Evolution and the Nature of Science.* Washington, DC: National Academy Press, 1998.

Nikaido, M., A. P. Rooney, and N. Okada. "Phylogenetic Relationships among Cetartiodactyls Based on Insertions of Short and Long Interspersed Elements: Hippopotamuses Are the Closest Extant Relatives of Whales." *Proceedings of the National Academy of Sciences, USA* 96 (1999): 10261–66.

Olby, R. "Francis Crick, DNA, and the Central Dogma." *Daedalus* 99 (1970): 938–87.

———. *Origins of Mendelism*. 2nd ed. Chicago: University of Chicago Press, 1985.

———. *The Path to the Double Helix*. Seattle: University of Washington Press, 1974.

Orel, V. *Gregor Mendel: The First Geneticist*. Oxford: Oxford University Press, 1996.

Otieno, A. C., A. B. Carter, D. J. Hedges, J. A. Walker, D. A. Ray, R. K. Garber, B. A. Anders, N. Stoilova, M. E. Laborde, J. D. Fowlkes, C. H. Huang, B. Perodeau, and M. A. Batzer. "Analysis of the *Alu* Ya-lineage." *Journal of Molecular Biology* 342 (2004): 109–18.

Paley, W. *Natural Theology*. London: J. Faulder, 1809.

Pearson, K. *Life, Letters, and Labours of Francis Galton*. Vol. 2. Cambridge: Cambridge University Press, 1924.

Perakh, M. *Unintelligent Design*. Amherst, NY: Prometheus Books, 2003.

Platinga, A. "When Faith and Reason Clash: Evolution and the Bible." *Christian Scholar's Review* 21, no. 1 (1991): 8–33.

Pollard, K. S., S. R. Salama, B. King, A. D. Kern, T. Dreszer, S. Katzman, A. Siepel, J. S. Pedersen, G. Bejerano, R. Baertsch, K. R. Rosenbloom, J. Kent, and D. Haussler. "Forces Shaping the Fastest Evolving Regions in the Human Genome." *PLoS Genetics* 2, no. 10 (2006): e168. Online at http://genetics.plosjournals.org.

Pollard, K. S., S. R. Salama, N. Lambert, M. A. Lambot, S. Coppens, J. S. Pedersen, S. Katzman, B. King, C. Onodera, A. Siepel, A. D. Kern, C. Dehay, H. Igel, M. Ares Jr., P. Vanderhaeghen, and D. Haussler. "An RNA Gene Expressed during Cortical Development Evolved Rapidly in Humans." *Nature* 443 (2006): 167–72.

Pope John Paul II. "Message to the Pontifical Academy of Sciences." *Quarterly Review of Biology* 72, no. 4 (1997): 381–83.

Portugal, F. H., and J. S. Cohen. *A Century of DNA: A History of the*

Structure and Function of the Genetic Substance. Cambridge, MA: MIT Press, 1977.

Proudfoot, N. J., A. Gil, and T. Maniatis. "The Structure of the Human Zeta-Globin Gene and a Closely Linked, Nearly Identical Pseudogene." *Cell* 31 (1982): 553–63.

Quintana-Murci, L., R. Chaix, R. S. Wells, D. M. Behar, H. Sayar, R. Scozzari, C. Rengo, N. Al-Zahery, O. Semino, A. S. Santachiara-Benerecetti, A. Coppa, Q. Ayub, A. Mohyuddin, C. Tyler-Smith, S. Qasim Mehdi, A. Torroni , and K. McElreavey. "Where West Meets East: The Complex mtDNA Landscape of the Southwest and Central Asian Corridor." *American Journal of Human Genetics* 74 (2004): 827–45.

Ray, J. *The Wisdom of God Manifested in the Works of Creation*. London: R. Harbin at the Prince's-Arms in St. Paul's Church Yard, 1717.

Redd, A. J., J. Roberts-Thomson, T. Karafet, M. Bamshed, L. B. Jorde, J. M. Naidu, B. Walsh, and M. F. Hammer. "Gene Flow from the Indian Subcontinent to Australia: Evidence from the Y Chromosome." *Current Biology* 12 (2002): 673–77.

Salem, A-H., D. Ray, J. Xing, P. A. Callinan, J. S. Myers, D. J. Hedges, R. K. Garber, D. J. Witherspoon, L. B. Jorde, and M. A. Batzer. "*Alu* Elements and Hominid Phylogenetics." *Proceedings of the National Academy of Sciences, USA* 100 (2003): 12787–91.

Sanger, F., A. R. Coulson, T. Friedmann, G. M. Air, B. G. Barrell, N. L. Brown, J. C. Fiddes, C. A. Hutchison III, P. M. Slocombe, and M. Smith. "Nucleotide Sequence of Bacteriophage Phi X174 DNA." *Nature* 265 (1977): 687–95.

Sanger, F., S. Nicklen, and A. R. Coulson. "DNA Sequencing with Chain-Terminating Inhibitors." *Proceedings of the National Academy of Sciences, USA* 74 (1977): 5463–67.

Sasagawa, I., M. Ishigooka, Y. Kubota, M. Tomaru, T. Hashimoto, and T. Nakada. "Pericentric Inversion of Chromosome 9 in Infertile Men." *International Urology and Nephrology* 30 (1998): 203–207.

Satta, Y., J. Klein, and N. Takahata. "DNA Archives and Our Nearest Relative: The Trichotomy Problem Revisited." *Molecular Phylogenetics and Evolution* 14 (2000): 259–75.

Sawada, I., C. Willard, C. K. Shen, B. Chapman, A. C. Wilson, and C.

W. Schmid. "Evolution of *Alu* Family Repeats since the Divergence of Human and Chimpanzee." *Journal of Molecular Evolution* 22 (1985): 316–22.

Scott, E. C. *Evolution vs. Creationism: An Introduction*. Westport, CT: Greenwood Press, 2004.

———. "Problem Concepts in Evolution: Cause, Purpose, Design, and Chance." National Center for Science Education, October 1, 1999. Online at http://www.ncseweb.org/resources/articles/ 695_problem_concepts_in_evolution_10_1_1999.asp (accessed December 30, 2006).

———. "NATB Statement on Evolution Evolves." Online at http:// www.ncseweb.org/resources/articles/8954_nabt_statement_on _evolution_ev_5_21_1998.asp (accessed November 24, 2005).

Seielstad, M., N. Yuldasheva, N. Singh, P. Underhill, P. Oefner, P. Shen, and R. S. Wells. "A Novel Y-chromosome Variant Puts an Upper Limit on the Timing of First Entry into the Americas." *American Journal of Human Genetics* 73 (2003): 700–705.

Shermer, M. S. *Why Darwin Matters: The Case against Intelligent Design*. New York: Times Books, Henry Holt & Co., 2006.

Shi, J., H. Xi, Y. Wang, C. Zhang, Z. Jiang, K. Zhang, Y. Shen, L. Jin, K. Zhang, W. Yuan, Y. Wang, J. Lin, Q. Hua, F. Wang, S. Xu, S. Ren, S. Xu, G. Zhao, Z. Chen, L. Jin, and W. Huang. "Divergence of the Genes on Human Chromosome 21 between Human and Other Hominoids and Variation of Substitution Rates among Transcription Units." *Proceedings of the National Academy of Sciences, USA* 100 (2003): 8331–36.

Shimamura, M., H. Yasue, K. Ohshima, H. Abe, H. Kato, T. Kishiro, M. Goto, I. Munechika, and N. Okada. "Molecular Evidence from Retroposons that Whales Form a Clade within Even-Toed Ungulates." *Nature* 388 (1997): 666–70.

Smith, C. M., and C. Sullivan. *The Top Ten Myths about Evolution*. Amherst, NY: Prometheus Books, 2007.

Spiegelman, S., B. D. Hall, and R. Storck. "The Occurrence of Natural DNA-RNA Complexes in *E. coli* Infected with T2." *Proceedings of the National Academy of Sciences, USA* 47 (1961): 1135–41.

Springer, M. S., A. Burk, J. R. Kavanagh, V. G. Waddell, and M. J.

Standhope. "The Interphotoreceptor Retinoid Binding Protein Gene in Therian Mammals: Implications for Higher Level Relationships and Evidence for Loss of Function in the Marsupial Mole." *Proceedings of the National Academy of Sciences, USA* 94 (1997): 13749–59.

Srebniak, M., A. Wawrzkiewicz, A. Wiczkowski, W. Kazmierczak, and A. Olejek. "Subfertile Couple with inv(2), inv(9) and 16qh+." *Journal of Applied Genetics* 45 (2004): 477–79.

Stern, C., and E. R. Sherwood. *The Origin of Genetics: A Mendel Source Book.* San Francisco: W. H. Freeman and Co., 1966.

Sturtevant, A. H. "A Crossover Reducer in *Drosophila Melanogaster* Due to Inversion of a Section of the Third Chromosome." *Biologisches Zentralblatt* 46 (1926): 697–702.

———. "The Effects of Unequal Crossing over at the *Bar* Locus in *Drosophila*." *Genetics* 10 (1925): 117–47.

———. *A History of Genetics.* New York: Harper and Row Publishers, 1965.

Szamalek, J. M., V. Goidts, D. N. Cooper, H. Hameister, and H. Kehrer-Sawatzki. "Characterization of the Human Lineage-Specific Pericentric Inversion that Distinguishes Human Chromosome 1 from the Homologous Chromosomes of the Great Apes." *Human Genetics* (2006) 120: 126–38.

Szamalek, J. M., V. Goidts, N. Chuzhanova, H. Hameister, D. N. Cooper, H. Kehrer-Sawatzki. "Molecular Characterisation of the Pericentric Inversion that Distinguishes Human Chromosome 5 from the Homologous Chimpanzee Chromosome." *Human Genetics* 117 (2005): 168–76.

Thomson, R., J. K. Pritchard, P. Shen, P. J. Oefner, and M. W. Feldman. "Recent Common Ancestry of Human Y Chromosomes: Evidence from DNA Sequence Data." *Proceedings of the National Academy of Sciences, USA* 97 (2000): 7360–65.

Torrents, D., M. Suyama, E. Zdobnov, and P. Bork. "A Genome-wide Survey of Human Pseudogenes." *Genome Research* 13 (2003): 2559–67.

Uddin, M., D. E. Wildman, G. Liu, W. Xu, R. M. Johnson, P. R. Hof, G. Kapatos, L. I. Grossman, and M. Goodman. "Sister Grouping of

Chimpanzees and Humans as Revealed by Genome-wide Phylogenetic Analysis of Brain Gene Expression Profiles." *Proceedings of the National Academy of Sciences, USA* 101 (2004): 2957–62.

Underhill, P. A., G. Passarino, A. A. Lin, P. Shen, M. Mirazon Lahr, R. A. Foley, P. J. Oefner, and L. L. Cavalli-Sforza. "The Phylogeography of Y Chromosome Binary Haplotypes and the Origins of Modern Human Populations." *Annals of Human Genetics* 65 (2001): 43–62.

Vigilant, L., M. Stoneking, H. Harpending, K. Hawkes, and A. C. Wilson. "African Populations and the Evolution of Human Mitochondrial DNA." *Science* 253 (1991): 1503–1507.

Volkin, E., and L. Astrachan. "Intracellular Distribution of Labeled Ribonucleic Acid after Phage Infection of *Escherichia coli*." *Virology* 2 (1956): 433–37.

———. "Phosphorus Incorporation in *Escherichia coli* Ribonucleic Acid after Infection with Bacteriophage T2." *Virology* 2 (1956): 149–61.

Wafaei, J. R., and F. Y. Choi. "Glucocerebrosidase Recombinant Allele: Molecular Evolution of the Glucocerebrosidase Gene and Pseudogene in Primates." *Blood Cells, Molecules, and Diseases* 35 (2005): 277–85.

Wang, H. Y., H. Tang, C. K. J. Shen, and C. I. Wu. "Rapidly Evolving Genes in Human. I. The Glycophorins and Their Possible Role in Evading Malaria Parasites." *Molecular Biology and Evolution* 20 (2003): 1795–1804.

Watson, J. D. *The Double Helix: A Personal Account of the Discovery of the Structure of DNA.* New York: W. W. Norton & Co., 1980.

Watson, J. D., and F. H. C. Crick. "A Structure for Deoxyribose Nucleic Acid." *Nature* 171 (1953): 737–38.

Wells, J. *Icons of Evolution: Why Much of What We Teach about Evolution Is Wrong.* Washington, DC: Regnery Publishing, 2002.

Wells, R. S., N. Yuldasheva, R. Ruzibakiev, P. A. Underhill, I. Evseeva, J. Blue-Smith, L. Jin, B. Su, R. Pitchappan, S. Shanmugalakshmi, K. Balakrishnan, M. Read, N. M. Pearson, T. Zerjal, M. T. Webster, I. Zholoshvili, E. Jamarjashvili, S. Gambarov, B. Nikbin, A. Dostiev, O. Aknazarov, P. Zalloua, I. Tsoy, M. Kitaev, M. Mirrakhimov, A.

Chariev, and W. F. Bodmer. "The Eurasian Heartland: A Continental Perspective on Y-chromosome Diversity." *Proceedings of the National Academy of Sciences, USA* 98 (2001): 10244–49.

Wells, S. *Deep Ancestry: Inside the Genographic Project.* Washington, DC: National Geographic Society, 2006.

Wheelwright, J. "Weaving the New Threads of Life." *Life* 3, no. 5 (1980): 48–55.

Whitcomb, J. C., and H. M. Morris. *The Genesis Flood: The Biblical Record and Its Scientific Implications.* Philadelphia: Presbyterian and Reformed Publishing Co., 1961.

White House, Office of the Press Secretary. "Text of Remarks on the Completion of the First Survey of the Entire Human Genome Project." Online at http://clinton5.nara.gov/WH/New/html/genome-20000626 .html (accessed January 27, 2007).

Wieland, C. "Darwin's Finches: Evidence Supporting Rapid Post-Flood Adaptation." *Creation* 14, no. 3 (1992): 22–23.

Wildman, D. E., M. Uddin, G. Liu, L. I. Grossman, and M. Goodman. "Implications of Natural Selection in Shaping 99.4% Nonsynonymous DNA Identity between Humans and Chimpanzees: Enlarging the Genus *Homo.*" *Proceedings of the National Academy of Sciences, USA* 100 (2003): 7181–88.

Wilkins, M. H. F. "Molecular Structure of Desoxypentose Nucleic Acids." *Nature* 171 (1953): 738–40.

Wilson, E. B. *The Cell in Development and Heredity.* 2nd ed. London: Macmillan, 1900.

Woese, C. R. *The Genetic Code: The Molecular Basis for Genetic Expression.* New York: Harper & Row Publishers, 1967.

Wong, W. C., J. F. Hess, C. K. Shen, B. Chapman, A. C. Wilson, and C. W. Schmid. "Comparison of Human and Chimpanzee Zeta 1 Globin Genes." *Journal of Molecular Evolution* 22 (1985): 309–15.

Wright, S. *Evolution and the Genetics of Populations.* Vols. 1–4. Chicago: University of Chicago Press, 1968–78.

———. "Evolution in Mendelian Populations." *Genetics* 16 (1931): 97–159.

Yang, F. B. Fu, P. C. O'Brien, W. Nie, O. A. Ryder, and M. A. Ferguson-Smith. "Refined Genome-wide Comparative Map of the Domestic

Horse, Donkey and Human Based on Cross-species Chromosome Painting: Insight into the Occasional Fertility of Mules." *Chromosome Research* 12 (2004): 65–76.

Yohn, C. T., Z. Jiang, S. D. McGrath, K. E. Hayden, P. Khaitovich, M. E. Johnson, M. Y. Eichler, J. D. McPherson, S. Zhao, S. Pääbo, and E. E. Eichler. "Lineage-Specific Expansions of Retroviral Insertions within the Genomes of African Great Apes but Not Humans and Orangutans." *PLoS Biology* 3, no. 4 (2005): e110. Online at http://biology.plosjournals.org.

Young, M., and T. Edis. *Why Intelligent Design Fails: A Scientific Critique of the New Creationism.* Rutgers, NJ: Rutgers University Press, 2004.

Yunis, J. J., and O. Prakash. "The Origin of Man: A Chromosomal Pictorial Legacy." *Science* 215 (1982): 1525–30.

Zerjal, T., R. S. Wells, N. Yuldasheva, R. Ruzibakiev, and C. Tyler-Smith. "A Genetic Landscape Reshaped by Recent Events: Y-chromosomal Insights into Central Asia." *American Journal of Human Genetics* 71 (2002): 466–82.

Zerjal, T., Y. Xue, G. Bertorelle, R. S. Wells, W. Bao, S. Zhu, R. Qamar, Q. Ayub, A. Mohyuddin, S. Fu, P. Li, N. Yuldasheva, R. Ruzibakiev, J. Xu, Q. Shu, R. Du, H. Yang, M. E. Hurles, E. Robinson, T. Gerelsaikhan, B. Dashnyam, S. Q. Mehdi, and C. Tyler-Smith. "The Genetic Legacy of the Mongols." *American Journal of Human Genetics* 72 (2003): 717–21.

Zhang, Z., and M. Gerstein. "The Human Genome Has 49 Cytochrome *c* Pseudogenes, Including a Relic of a Primordial Gene that Still Functions in Mouse." *Gene* 12 (2003): 61–72.

INDEX